Lagrangian and Hamiltonian Mechanics

José Rachid Mohallem

Lagrangian and Hamiltonian Mechanics

A Modern Approach with Core Principles and Underlying Topics

 Springer

José Rachid Mohallem
Physics
Universidade Federal de Minas Gerais
Belo Horizonte, Brazil

ISBN 978-3-031-55204-5   ISBN 978-3-031-55202-1   (eBook)
https://doi.org/10.1007/978-3-031-55202-1

This Springer imprint is published by the registered company Springer Nature Switzerland AG
The registered company address is: Gewerbestrasse 11, 6330 Cham, Switzerland

Paper in this product is recyclable.

Preface

In the beginning, there was the Action Principle.

This book was originally intended to be used as a complementary text to available textbooks of Analytical Mechanics, involving topics whose development lacks due clarity in these texts. It results from author's observations on recurring difficulties from students and especially focuses on the more obscure theoretical issues that generate these difficulties. It is therefore appropriate for complementary use by lecturers and for student learning. In the course of writing, however, it was transformed into a text adaptable to a semester course in Analytical Mechanics, so that an almost complete chapter on Hamiltonian mechanics is included. Nevertheless, the reader will note that its central aim is exploring the outstanding idea that searching Lagrangians from symmetries and making use of Hamilton's principle is probably the more powerful theoretical instrument of physics.

The theory of Classical Mechanics, that is, the study of non-quantum motions of particles and bodies, can be presented in different ways, according to the objective of each course. For the future researcher in physics, less than practical applications, the major interest lies in the methods of mechanics, which constitute the source and basis of concepts for the development of more advanced theories, such as Statistical Mechanics, Quantum Mechanics and Field Theories. These methods are wonderfully illustrative of how research in theoretical physics works. This aspect, I believe, is one of the most flawed in the (yet excellent) texts commonly used, which privilege a supposedly logical and deductive construction of Mechanics. A major advantage of the analytic formulation over Newtonian mechanics is exactly that of making possible to illustrate the inductive character of any theory under construction about how nature works.

For example, despite its enormous importance, expressing the Lagrangian of a classical conservative system as $L = T - V$ right from the start, relegating the other cases to the merely curiosities, drives the student thinking that this would be the "normal" form of the Lagrangian. The very Hamilton's principle is then often interpreted in a wrong way, that is, as an alternative way of obtaining the results of Newtonian Mechanics. Its physical content, which involves a wide arbitrariness in the expression of the Lagrangian, is often overlooked. Another source of misinterpretations is the presentation of the Lagrangian theory, even if starting from Hamilton's principle, exclusively in generalized coordinates, which is a *sine qua non* condition only to the old formulation based on D'Alembert's principle. The elimination of forces of constraints, objective of this principle, has long ceased to be the motivating factor for Lagrangian mechanics, but strangely, some books still highlight the "symmetry" of the two possible ways to arrive to Lagrange's equations.

Another characteristic in the basic literature of analytical mechanics is the almost absolute lack of discussion of the difficulties, but also of the revelations, introduced by the presence of moving constraints. This is a sensitive point in learning, since it leads to recurring mistakes, especially regarding the conservation of the the the Jacobi integral h and/or the equality $h = E$, as well as in the interpretation of common terms of the equations of motion in an inertial frame of reference and in another one associated to the constraint, the last possibly non-inertial.

Although the examples cited above are central to this text, many other less clear aspects of the theory are also focused, aiming at accounting to recurring student queries. The elucidation of the parametric role of time in the classical and non-covariant relativistic formulations; of the position and the velocity (or momentum) as independent variables, as well as the implicit and explicit dependencies between dynamic variables, aim to eliminate difficulties that refer back to the basic courses of calculus.

Further contributions addressed are: (1) the proper understanding of how Lagrangian (2nd order) and Hamiltonian (1st order) equations work in practice; (2) the elimination of the unnecessary reference to sympletic notation in the proof of invariance of the Poisson square brackets; (3) when to choose Euler's angles as generalized coordinates (instead of using Euler's vector equations) in the description of the dynamics of a rigid-body; (4) recognizing the importance and difficulties about the real nature of adiabatic invariants. Last but not least, (5) the fundamental topic of Galilean invariance, together with the identification of where Newton's first and third laws lie in Lagrange mechanics, stand out as unsure points that deserve much more attention in a textbook.

The reader will understand that the organization and the choice of themes for this text aims to fill these and other gaps detected by the author in years teaching this discipline. On the contrary of being simply a critique of existing approaches, it is intended to complement them, in order to make explicit the potential of the theory in its actual magnitude.

In order to use this book, it is assumed, on the part of a student, the knowledge of basic Newtonian theory at undergraduate level and rudiments of Lagrangian and Hamiltonian mechanics, electromagnetism and special theory of relativity.

The author is in debt to colleagues and students of the Department of Physics of UFMG for encouragement and enlightening discussions. Particular thanks are due: To professor Wagner Corradi, coordinator of the open university project, and technical staff, for publishing an old version of the book. To professor Emmanuel Pereira for very helpful discussions. Particularly, to professors Glauber Dorsch and Carlos Heitor Fonseca for their invaluable help with classical field theory and relativity, respectively. It is not possible to name all the students who contributed criticisms, comments and suggestions, without the risk of leaving many of them out, but everyone can be assured that their help was also invaluable. Last but not least, to Nemul Khan and Nivetha Moorthi, the Springer editors assigned to this publication, for their enthusiasm and support, and to my dear daughter Tássia D. S. Mohallem for drawing the nice figures.

Belo Horizonte, Brazil José Rachid Mohallem

Contents

Chapter 1
Concepts and Principles

Abstract After presenting the basic concepts to start the study of mechanics, such as time, space, particle and mass, a necessary mathematical review of implicit and explicit dependence of variables and variational calculus is made. Then, Hamilton's principle is introduced as the common origin of all phenomena in classical mechanics. The discovery of the form of the Lagrangian function and its dependence on dynamic variables, to be used with Hamilton's principle, is presented as the true theoretical challenge, exemplifying the inductive character of physical theories. Several examples of Lagrangian functions different of the "standard" form, $L = T - V$, are discussed for a particle free of constraints, that is, prior to the introduction of generalized coordinates and the $L = T - V$ form. Obtaining Newton laws from Hamilton's principle and symmetries of nature, reinforces the basic idea of the chapter, that is, the prevalence of Hamilton's principle, which works even in cases where Newton's laws do not, as in special relativity.

Keywords Hamilton's principle · Symmetries · "Magic" Lagrangians · Generalized potentials · Equivalent Lagrangians · Newton's laws · Connection to quantum mechanics

1.1 Fundamentals

Classical Mechanics is developed within the 3D (three-dimensional) space as its scenario, space and time taken as intuitive and independent concepts. In this scenario, the actors are mass point particles and their interactions, described by forces or fields. Extensive bodies can be represented by sets of rigidly bound particles, or even by rigid continuous mass distributions. Less often it is necessary to consider amorphous mass distributions, as in mechanical theories of fluids or elastic bodies. Body movements are often sufficiently well described by the translation of a "particle"—its center of mass (CM)—associated with its total mass and, if necessary, by rigid rotations.

The mass is taken as an intrinsic attribution to the particles, a property that characterizes their reaction to interactions, that is, their inertia, as well as the intensity of a specific interaction, the gravitational one. Other property, the electric charge, characterizes the intensity of the electromagnetic interaction, in cases where classical mechanics is appropriate to describe the phenomena involved. Among the fundamental forces of nature, the gravitational and electromagnetic ones are the only that fit to a classical description.

It is not always possible to model mechanical systems of interest with only these elements, however. Static friction, for example, among other surface forces, comes from the action of microscopic forces, which, in addition to existing in extraordinarily large number, do not individually obey the laws of classic mechanics. Empirical observations lead to the inclusion of these effects through forces that geometrically or kinematically constrain the motion of the studied particles. These are the constraint forces, which enter the theory as unknowns (normal reaction force, tension force etc.). Most constraint forces, the "geometric ones", constrains the configurational space of movement. In contrast, kinematic constraint forces restrict the allowable speeds. Dynamic friction forces (e.g. friction in fluids) also resulting from an uncountable number of interactions at the atomic level, need empirical expressions independent of the solution of the dynamical problem, since they also determine the dynamical behavior of the system. It is correct practice to consider kinematic friction, even surface friction, an empirically described dynamical agent.

In general, classical mechanics does not admit physical considerations on the microscopic level, although it is, in some aspects, a point of passage for the formalism of quantum mechanics. This can be done either in the context of Lagrangian mechanics, through Feynman's path integrals, or in the context of Hamiltonian mechanics. However, this is achieved through drastic postulations, whereby paradigms vital to classical mechanics, such as the determinism of movements, are replaced by new paradigms.

In the case of special relativistic theory, the situation is different. The non-covariant formulation of the theory of special relativity has a very close relationship with Newtonian mechanics. Through the freedom of choice of the Lagrangian function for Hamilton's principle, non-covariant special relativity can be treated within the same theoretical frame that yields Newtonian mechanics from its Lagrangian counterpart. Even the covariant formulation, in which it is rather necessary to consider the four-dimensional space-time as a new scenario, is subject to an appropriate Lagrangian formulation.

Finally, a few words about epistemological terminology. A physical *principle* is a non demonstrable proposition taken from the beginning, provisionally unquestionable. A physical *law* arises from repetitive observations of a regular behaviour in nature. The term is less rigid, but can be accepted as being also a principle, once the law is accepted as (provisionally) unquestionable. A *theorem* is a proposition that needs to be demonstrated from fundamental principles. As examples we have, respectively, Hamilton's principle, the law of conservation of energy and Nöther's theorem.

1.1.1 Dynamical Variables, Explicit and Implicit Dependences

In classical mechanics, the study of the movement of one (or several) particle assumes time t and space as intuitive and absolute existing quantities. Particularly, t parametrizes the evolution of a dynamical system, in the sense of changing of the positions of its particles as this parameter increases. The time dependence of dynamic quantities is a priori unknown. In fact, it is the task of dynamics to determine this dependence through the equations of motion. On the other hand, the time t can appear explicitly in the equations of motion in other ways, for example, through interaction of the particle with variable fields or through moving constraints, as will be seen. The time t, taken in this sense, intuitive and independent of any other concept, also allows the non-covariant treatment of special relativity, but not the space-time covariant treatment. This means that the non-covariant relativistic equations are valid in the specific inertial frame we work with.

For notational simplicity, we refer here to a single particle in 1D (one-dimensional) motion along the x axis. The very definitions of kinematic quantities, velocity and acceleration, are based on our intuition about space and time. Formally, the travelled (in space) distance x is considered a function of t, $x(t)$, even this dependence being unknown prior to the solution of the equation of motion. Velocity $\dot{x}(t) = \dfrac{dx}{dt}$ is also a function of t, as is acceleration $\ddot{x}(t)$. Returning temporarily to the 3D space, it seems clear that, in order to describe a possibly curved particle's trajectory, we need to know where it is but also where it is going to at each instant t, that is, we need $\overrightarrow{r}(t)$ and $\overrightarrow{v}(t)$.

It turns out that the variables $x(t)$ and $\dot{x}(t)$ are necessary to determine the trajectory of the particle in space and are, therefore, fundamental variables in the description of the motion (recalling that the equation of motion is a second-order differential equation, containing a term in $\ddot{x}$). In what follows, it will be convenient to consider t as our *independent variable* and $x(t)$ and $\dot{x}(t) = \dfrac{dx}{dt}$ as *fundamental* variables, independent of each other, whose dependence on t is previously unknown.

In consequence, the trajectory of the particle in the "phase space",[1] $\dot{x}$ versus x, is such that each of its points represents the mechanical *state* of the particle.

So, x and $\dot{x}$, despite the definition $\dot{x} = \dfrac{dx}{dt}$, can be treated as *explicitly independent* (among themselves) variables, even on the physical trajectory. This is true because the elimination of t from the set $x = x(t)$ and $\dot{x} = \dot{x}(t)$, whenever possible, yields only an *implicit* dependence $\dot{x} = \dot{x}(x)$. Perhaps the simplest and clearest example is the uniform motion, where $\dot{x} = \dot{x}_0$, constant, while $x = \dot{x}_0 t$, so that t cannot even be eliminated. For each t, $\dot{x}$ remains the same, while x varies linearly with time.

---

[1] This denomination is used in introductory textbooks but the "real" phase space will be defined later.

Despite that, the previous definition $\dot{x} = \dfrac{dx}{dt}$ establishes a kind of *previous* connection between the two variables, which will be reinterpreted in Hamiltonian mechanics in order to generate fully independent variables. This matter is treated in Chap. 3.

Let us go a little further on this subject. Consider an implicit dependence $\dot{x} = \dot{x}(x)$ as,

$$f(\dot{x}, x) = 0, \tag{1.1}$$

obtained after solving the equations of motion and eliminating t as above. As examples, we have $\dot{x}^2 - \dot{x}_0^2 - 2ax = 0$, in rectilinear uniform motion or $\dot{x}^2 + \omega^2 x^2 - 2E/m = 0$ for the 1D harmonic oscillator, in standard notation. In these two cases, the equations can be also obtained by using energy conservation. It's never too much to repeat that these dependences are implicit, there should be no doubt about this. The derivative $\dfrac{d\dot{x}}{dx}$, for example, can be obtained through implicit derivation of Eq. (1.1). On the other hand, the partial derivative $\dfrac{\partial \dot{x}}{\partial x}$ does not exist, or else, it is equal to $\dfrac{d\dot{x}}{dx}$.

The concepts of implicit and explicit dependence of variables, particularly with t, are of such importance in mechanics that they deserve greater detailing.

The time derivatives of a quantity $A(x, t)$ are mathematically defined as

$$\frac{\partial A}{\partial t} = \lim_{\Delta t \to 0} \frac{A(x, t + \Delta t) - A(x, t)}{\Delta t} \tag{1.2}$$

and

$$\frac{dA}{dt} = \lim_{\Delta t \to 0} \frac{A[x(t + \Delta t), t + \Delta t] - A(x, t)}{\Delta t}. \tag{1.3}$$

We understand the partial derivative above as the evaluation of the temporal variation of A for fixed values of x, while the total derivative evaluates its unrestricted variation with t.

So, let us take a function $A(x, t)$ in two situations,

$$A(x, t) = f(x), \text{ and } x = x(t) \text{ (implicit in t)} \Rightarrow \frac{\partial A}{\partial t} = 0, \quad \frac{dA}{dt} \neq 0 \tag{1.4}$$

$$A(x, t) = f(x, t), \text{ and } x = x(t) \text{ (explicit in both)} \Rightarrow \frac{\partial A}{\partial t} \neq 0, \quad \frac{dA}{dt} \neq 0$$

In the first case, we say that A depends explicitly on x and implicitly on t; that is, if x is fixed, A does not change in time. In the second, it depends explicitly on both variables, for example, $A = x^2 + sin(t)$ or $A = e^t x$.

If A depends on several variables $x_i(t)$, the total derivative is

$$\frac{dA}{dt} = \sum_i \frac{\partial A}{\partial x_i}\dot{x}_i + \frac{\partial A}{\partial t}.$$ (1.5)

Back to one dimension, since x e $\dot{x}$ depend exclusively on t, it becomes evident that, for them

$$\frac{\partial x}{\partial t} \equiv \frac{dx}{dt}, \quad \frac{\partial \dot{x}}{\partial t} \equiv \frac{d\dot{x}}{dt}.$$ (1.6)

Other quantities are written as functions of the variables x and $\dot{x}$. Let us take the example of the mechanical energy of a particle in simple harmonic motion.

$$E = \frac{m}{2}\dot{x}^2 + \frac{k}{2}x^2.$$ (1.7)

We see that E depends explicitly on x and $\dot{x}$ and, through them, implicitly on t. Formally,

$$\frac{dE}{dt} = \frac{\partial E}{\partial \dot{x}}\ddot{x} + \frac{\partial E}{\partial x}\dot{x}, \qquad \frac{\partial E}{\partial t} = 0.$$ (1.8)

The other derivatives are

$$\frac{dE}{dx} = \frac{\partial E}{\partial \dot{x}}\frac{d\dot{x}}{dx} + \frac{\partial E}{\partial x}; \quad \frac{dE}{d\dot{x}} = \frac{\partial E}{\partial x}\frac{dx}{d\dot{x}} + \frac{\partial E}{\partial \dot{x}},$$ (1.9)

where the existence of the last partial derivatives accounts for the explicit dependence of E on x and $\dot{x}$. Let us now consider that the only force that acts on the particle is the harmonic force. We know that, in this case, mechanical energy will be conserved, that is, any of the total derivatives above will vanish. This is proved with the help of equation of motion $m\ddot{x} + kx = 0$. In fact,

$$\frac{dE}{dt} = m\dot{x}\ddot{x} + kx\dot{x} = \dot{x}(m\ddot{x} + kx) = 0;$$ (1.10)

$$\frac{dE}{dx} = m\dot{x}\frac{d\dot{x}}{dx} + kx = m\frac{d\dot{x}}{dt} + kx = 0;$$

$$\frac{dE}{d\dot{x}} = kx\frac{dx}{d\dot{x}} + m\dot{x} = \frac{dx}{d\dot{x}}(kx + m\frac{dx}{dt}\frac{d\dot{x}}{dx}) = 0.$$

Therefore, the explicit dependence of E on x, for example, only guarantees the existence of the partial derivative $\dfrac{\partial E}{\partial x}$, the total derivative being null. The reader must then realize that the energy E, despite being constant, is a function of x and $\dot{x}$, according to Eq. (1.7). In other words, the fact that E is constant, even null, does not invalidate its dependence on x and on $\dot{x}$, a feature that will be often highlighted in future developments. Indeed, the constant value of E depends on the initial conditions $x(t = 0)$ and $\dot{x}(t = 0)$. Finally, in the case of two dimensions,

$$E = \frac{m}{2}(\dot{x}^2 + \dot{y}^2) + \frac{k}{2}(x^2 + y^2), \tag{1.11}$$

the derivatives can be obtained from the differential of E,

$$dE = \frac{\partial E}{\partial \dot{x}}d\dot{x} + \frac{\partial E}{\partial \dot{y}}d\dot{y} + \frac{\partial E}{\partial x}dx + \frac{\partial E}{\partial y}dy. \tag{1.12}$$

For example,

$$\frac{dE}{dx} = \frac{\partial E}{\partial \dot{x}}\frac{d\dot{x}}{dx} + \frac{\partial E}{\partial \dot{y}}\frac{d\dot{y}}{dx} + \frac{\partial E}{\partial x} + \frac{\partial E}{\partial y}\frac{dy}{dx}, \tag{1.13}$$

which means again that a derivative like $\dfrac{d\dot{x}}{dx}$ can exist and be obtained from this expression. On the other hand, derivatives like $\dfrac{\partial \dot{x}}{\partial y}$ or $\dfrac{\partial y}{\partial x}$ do not exist, as these variables are independent between each other.

Note that the last term of Eq. (1.13) contains the derivative $\dfrac{dy}{dx}$. At this point this should be no surprise, since y and x are connected through the equation of trajectory in the coordinate space. That is, the derivative exists *on the trajectory*. Otherwise, x and y are fully independent variables so that it makes no sense to consider $\dfrac{dy}{dx}$ or $\dfrac{\partial y}{\partial x}$ out of the trajectory.

It should also be noted, finally, that the explicit independence of E relative to t does not necessarily imply its conservation. For example, if there is a dissipative force $-c\dot{x}$ acting on the oscillator, the first of Eqs. (1.10) changes to

$$\frac{dE}{dt} = \dot{x}(m\ddot{x} + c\dot{x} + kx) - c\dot{x}^2 = -c\dot{x}^2, \tag{1.14}$$

because now it is the term in parentheses that represents the equation of motion and cancels out. In this case, E is not conserved, but it is still explicitly independent on time. Nonetheless, it is possible that E depends explicitly on t in particular cases, as, for example, that of a time-dependent potential energy (a charged particle in an oscillating electric field, for instance); E will not be conserved and we will have $\dfrac{dE}{dt} \neq 0$ and also $\dfrac{\partial E}{\partial t} \neq 0$.

What was stated in this section is valid for analogous quantities: vectors, curvilinear coordinates, etc., for one or several particles. In Hamiltonian Mechanics, instead of velocities $\dot{x}$, the quantities used are the so-called generalized momenta p, with no relevant changes relative to the above statements.

1.2   Calculus of Variations

Consider the following question: If a projectile, in the Earth gravitational field $\vec{g}$, passes through the point (x_1, y_1) at instant t_1 and the point (x_2, y_2) at instant t_2, what is the path followed by it between t_1 and t_2? Since we are used to Newtonian mechanics, by which we give the projectile's position and velocity in t_1 and ask where and with what velocity it will be in future times, t_2 for instance (Fig. 1.1), the question posed above is commonly viewed with strangeness at first sight. What usually comes to mind is: how do we know where it will be in t_2, without having "followed" its motion from t_1 to t_2, by solving Newton's equation of motion? We will see now that this doubt is unjustified.

Assembling the well-known parametric equations of the trajectory, as functions of the parameter t, from Newtonian mechanics, we have

$$x_2 = x_1 + v_{x1}(t_2 - t_1), \qquad y_2 = y_1 + v_{y1}(t_2 - t_1) - \frac{1}{2}g(t_2 - t_1)^2, \quad (1.15)$$

$$v_{x2} = v_{x1}, \qquad v_{y2} = v_{y1} - g(t_2 - t_1).$$

The four equations above have, in principle, eight unknowns: x_1, y_1, v_{x1}, v_{y1} (in t_1) and x_2, y_2, v_{x2}, v_{y2} (in t_2). Four of them, usually the first set (the "initial conditions") must be specified, in order to obtain the second set through Eqs. (1.15).

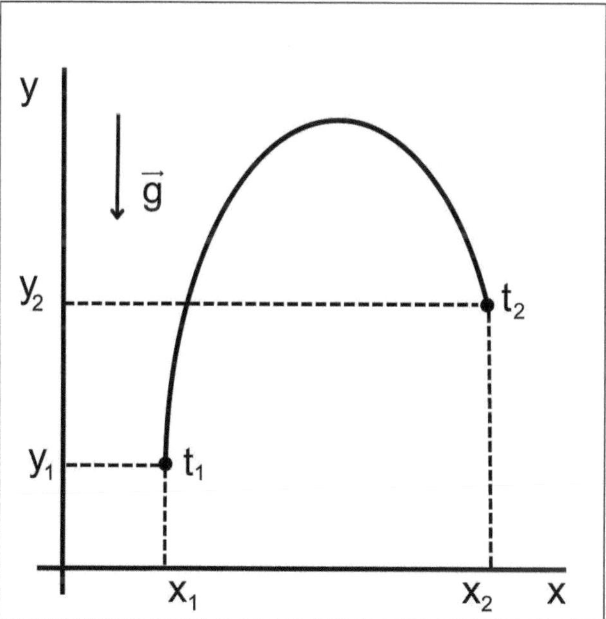

Fig. 1.1 The initial and final positions define the trajectory of a projectile

However, it is mathematically equivalent to specify the four position coordinates, x_1, y_1, x_2, y_2 (the"boundary conditions") and then obtain, from Eqs. (1.15), the four velocity components at the two instants, v_{x1}, v_{y1}, v_{x2}, v_{y2}. The doubt posed above is therefore unwarranted. In other words, the initial and final positions specify unequivocally the particle's trajectory and we need just to get used to the new paradigm.

In Newtonian mechanics, we provide the initial conditions and build the parametric trajectory point by point as t evolves (i.e., for every dt), by solving the differential equations of motion. An illustration of this procedure might be given by a numerical solution of this equations. But for "jumping", directly from t_1 to t_2, we need a new mathematical tool, the Calculus of Variations. We can pose the problem of variational calculus, aiming already at its mechanical applications, as follows:

Given a *functional* $J[\vec{r}(t)]$, which is a quantity dependent on an indeterminate function $\vec{r}(t)$, *for which* $\vec{r}(t)$ *will J be a minimum?*[2]

With no loss of generality, let us save notation by choosing just the $x(t)$ component of $\vec{r}$ for the next derivation. Producing a numerical quantity from a function can be done in countless ways; let us be more specific, restricting the variational problem to what will actually interest us here, taking

$$J[x(t)] = \int_{t_1}^{t_2} f(x, \dot{x}, t)dt, \tag{1.16}$$

where the points $x_1(t_1)$ and $x_2(t_2)$ are fixed. Note then that we establish the functional as an definite integral, *a number*, dependent on the function $x(t)$. The specification of $f(x, \dot{x}, t)$ can be postponed, since the minimization of J will be done relative to $x(t)$, regardless of the form of f.

Unlike the usual maximum and minimum calculations, in which we would vary t to find an extremum of $x(t)$, here we want to vary a function $x(t)$ between t_1 and t_2 to find an extremum of J. Infinite possibilities exist for $x(t)$, see Fig. 1.2. Let us consider a generic functional variation relative to the actual solution $x(t)$, such that, *for each value of* t, we have

$$\bar{x}(t) = x(t) + \delta x(t), \tag{1.17}$$

where $\delta x(t)$ is a continuous and differentiable function that vanishes at the extremes, $\delta x(t_1) = \delta x(t_2) = 0$. The reader may wonder what it means to vary a function of t at fixed values of time, but there is, in fact, no difficulty with this concept. As shown

---

[2] The reason to choose a minimum, not a extremum, will be presented later, but for the present development, an extremum is fine.

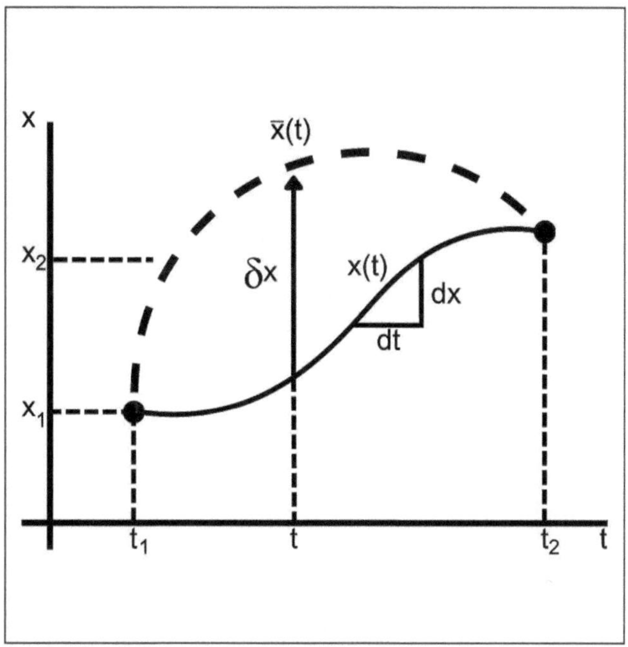

Fig. 1.2 The functional variation $\delta x(t)$ of parametric trajectories of a particle. Note that $\delta x(t)$ is made at fixed t, while the usual differential dx taken over the trajectory implies on a corresponding variation dt

in Fig. 1.2, for a given value of t, the value of the varied function $\bar{x}(t)$ is obtained by adding to $x(t)$ the variation $\delta x(t)$ at the same instant. Equivalently, the variation $\delta x(t)$ can be seen as the difference between $\bar{x}(t)$ and $x(t)$ at the instant t.

The corresponding variation in the functional will be

$$\delta J[x(t)] = \delta \int_{t_t}^{t_2} f(x, \dot{x}; t)dt = \int_{t_t}^{t_2} \delta f(x, \dot{x}; t)dt = \int_{t_t}^{t_2} [\frac{\partial f}{\partial x}\delta x + \frac{\partial f}{\partial \dot{x}}\delta \dot{x}]dt,$$

$$(1.18)$$

where $\dfrac{\partial f}{\partial t} = 0$, since the variations are made with fixed t. The passage of δ to inside the integral is granted by the extremal points being fixed. Now note that $\delta \dot{x} \equiv \delta \dfrac{dx}{dt} = \dfrac{d}{dt}\delta x$, since the functional variation, being made at fixed t, commutes with $\dfrac{d}{dt}$. So,

$$\delta J[x(t)] = \int_{t_t}^{t_2} [\frac{\partial f}{\partial x}\delta x + \frac{\partial f}{\partial \dot{x}}\frac{d}{dt}\delta x]dt.$$

$$(1.19)$$

The integral of the second term, performed by parts ($\int u\,dv = uv - \int v\,du$ with $u = \dfrac{\partial f}{\partial \dot{x}}$ and $dv = d(\delta x)$), produces a null term $\left[\dfrac{\partial f}{\partial \dot{x}}\delta x\right]_{t_1}^{t_2} = 0$, since δx vanishes at the extreme points. We are then lead to

$$\delta J[x(t)] = \int_{t_t}^{t_2} [\frac{\partial f}{\partial x} - \frac{d}{dt}\frac{\partial f}{\partial \dot{x}}]\delta x\,dt. \tag{1.20}$$

The condition that $J[x(t)]$ is an extreme, $\delta J[x(t)] = 0$, combined with the arbitrariness of the variation δx, leads finally to

$$\frac{\partial f}{\partial x} - \frac{d}{dt}\frac{\partial f}{\partial \dot{x}} = 0. \tag{1.21}$$

This differential equation, called the Euler-Lagrange (or simply Lagrange) equation, must be obeyed so that $x(t)$ minimizes the functional J. Several mathematical or geometrical problems are posed in a form analogous to the minimization of the functional (1.16). Here, we restrict ourselves to cases in which the independent variable is time t, which conducts us directly to mechanical problems.

Methods dealing with search of extrema in the above or an analogous form are usually called variational methods. We will later discuss the corresponding problem in more dimensions, in a context of greatest interest, after the introduction of the generalized coordinates.

Finally, we should point out that, once the function $f(x, \dot{x}, t)$ is known, the minimization of the functional J is equivalent to the solution of the Euler-Lagrange equation (1.21). Adding equivalent derivations for the other independent components of $\vec{r}$, the same Eq. (1.21) will be obeyed in vector form.

1.3   Hamilton's Principle

In this section, we will see how the proper choice of a function of coordinates, velocities and time, the *Lagrangian* of a particle, playing the role of the function f, yields the proper equations of motion as a variational principle, *Hamilton's principle*, is invoked:

For a moving particle, there is a function called Lagrangian,

$$L(\vec{r}, \dot{\vec{r}}, t) \tag{1.22}$$

such that the action functional is defined by

$$S = \int_{t_1}^{t_2} L(\vec{r}, \dot{\vec{r}}, t)\,dt. \tag{1.23}$$

The reason for the Lagrangian to depend on $\vec{r}$ and $\dot{\vec{r}}$ can be referred back to Newtonian kinematics. Since we search for the particle's trajectory, which depends on our knowledge of where the particle is but also where it is going to (even in 1D there are two possibilities), it is expected that L should depend on $\vec{r}$ and $\dot{\vec{r}}$. In fact, the *explicit* dependence on t should be more surprising. Later we will discuss problems involving moving constraints, but for now it is enough to realize that the potential energy might be a function of t. We will soon also justify, in more solid physical grounds, the arguments of L.

Once we agree with $L = L(\vec{r}, \dot{\vec{r}}, t)$, we are ready to stablish Hamilton's principle:

Among all possible trajectories between the instants t_1 and t_2, with $\vec{r}(t_1)$ e $\vec{r}(t_2)$ fixed, the actual trajectory of the particle will be that which minimizes the action, that is,

$$\delta S = 0. \tag{1.24}$$

According to the derivation leading to Eq. (1.21) the famous Lagrange's equations are then,

$$\frac{\partial L}{\partial x_i} - \frac{d}{dt}\frac{\partial L}{\partial \dot{x}_i} = 0, \quad i = x, y, z. \tag{1.25}$$

(check it!)

Since the term "action" has long been used in the literature to mean different quantities, we avoid referring to the stated principle as "principle of least action", preferring the use of "Hamilton's principle".

Note that the functional dependence of L with its arguments is still unknown. In fact, the choice of this dependence is as fundamental as Hamilton's principle for mechanics. The are at the very hearth of the theory.

Before moving on to applications to real motions, some last points must be stressed. From Newtonian mechanics, we know that the future motion of a classical system is determined as we obtain the positions and velocities of its particles in function of t, by solving the equations of motion of each particle α, $\vec{F}_\alpha = m_i \dfrac{d\vec{v}_\alpha}{dt}$, provided their initial values are known. This led us to consider, in Sect. 1.1.1, x_i and $\dot{x}_i$ as functions of t independent of each other. This actual independence, assumed in the construction of the Lagrangian, does not find, however, a clear formal correspondence in theory, since there is an a priori relationship $\dot{x}_i = \dfrac{dx_i}{dt}$. This apparent incongruity does not actually lead to troubles, because nothing exists in the Lagrangian formalism that requires the independent treatment of these variables. In fact, we got the Euler-Lagrange equation (which corresponds, as the reader may have already suspect, to the equation the motion) from Hamilton's principle, which demands the variation of only $x(t)$ and not $\dot{x}(t)$. The variable $\dot{x}$ is involved in the Lagrangian by postulation, while the Euler-Lagrange equation, being a second-order differential equation on $x(t)$, automatically involves $\dot{x}(t)$ as part of its solution. When we introduce the Hamiltonian mechanics, based on first-order equations of motion for each variable, we will see that $\dot{x}$ will not be the convenient variable to pair with x. Instead, it will be replaced by the so-called generalized momenta.

Exercise Consider a particle of mass m in 1D motion under gravity, so that $V = mgy$. At $t = 0$, $y = 0$. At $t = \tau$, $y = H$. Consider $L = T - V$. (a) Using the family of functions $y(t) = At^2 + Bt + C$, show that $C = 0$ and $B = \dfrac{H}{\tau} - A\tau$, as you use the boundary conditions. (b) Show that $\frac{\partial S}{\partial A} = 0$ yields the correct expression $y(t) = -\frac{1}{2}gt^2$. (c) Now, try to repeat the procedure but with the family $y(t) = At^2$ simply, suggested by the previous solution and according to the initial conditions. You should find wrong multiple results, see [1].

The conclusion is that a problem in the form of Hamilton's principle is a problem of *boundary conditions*, not of *initial conditions*. However, this feature is already accounted for when Lagrange's equations are deduced, so that their use is safe without such concerns.

Last but not least, the general principle of relativity states that the laws of physics are invariant under exchange of inertial reference frames. Since the action S must be an escalar, it must be invariant under such changes. In the scope of Newtonian mechanics, this means that they must be invariant under Galilean transformations, but in the scope of Einstein's relativity, they must be invariant under Lorentz' transformations. We will have opportunity to explore these features in many times as we go forward.

1.4   Constructing Lagrangians from Symmetries

The search for forms of Lagrangians is inductive and must be based on fundamental symmetry properties of space and time. In order to gain insights, we start with the simplest system, a free particle with mass m in Newtonian mechanics. What is the form of L in this case? Note first that L can not depend on x, since all points in space must be equivalent for a free particle. In other words, *empty space is homogeneous*. Since *time is also homogeneous*, L does not depend explicitly on t as well. So, L can depend just on the particle's velocity $\vec{\dot{r}}$. But *space is also isotropic*, no directions are preferable (the origin of the coordinate system is also immaterial), hence L can depend only on $|\vec{\dot{r}}| = v$.

What should be the exact form of this dependence? In this book we adopt the point of view that Hamilton's principle has priority over specific forms of coordinate and velocity transformations, which must stem from its application. Let us consider $L = Av^2$. Instead of taking it into Lagrange's equation (1.21), it seems more instructive to use Hamilton's principle from the beginning.

For the action integral we have, $S = \int_{t_1}^{t_2} A\dot{x}^2 dt$, so that, $\delta S = 2A \int_{t_1}^{t_2} \dot{x}\delta\dot{x}dt = 2A \int_{t_1}^{t_2} \dot{x}d(\delta x)$. Integrating by parts, as usual, we get $\delta S = [2A\dot{x}\delta x]_{t_1}^{t_2} - 2A \int_{t_1}^{t_2} \delta x \frac{d\dot{x}}{dt}dt$ where, in the last part, we used the commutation of the operators d and δ. In turn, the integral of the first term above vanishes as we know, so that $\delta S = -2A \int_{t_1}^{t_2} \ddot{x}\delta x dt$.

As we impose $\delta S = 0$, being δx arbitrary, we are lead to $\ddot{x} = 0$, an uniform motion, as expected. It may be not so clear at this point, but latter we will generalize this procedure and show that the trial Lagrangian is searched in order to be invariant under a Galilean transformation of velocities.

Exercise Try $L = Av$ (you should have a negative answer).

So, we assume $L = Av^2$ and latter show that this choice allows, in general, obtaining Newton's first law, which is presently taken as defining an inertial frame.

What about the proportionality constant A? Once we expect Lagrange's equation to repeat Newton's second law[3] $m\ddot{x} = 0$, then we must have $A = \frac{1}{2}m$, which leads L to be identified with the kinetic energy of the particle,

$$L = T = \frac{1}{2}m\dot{x}^2. \tag{1.26}$$

Here we find the reason to call Hamilton's principle as principle of least (or minimal) action. Let $\dot{x} = v_0$ or $x = v_0 t$. Since it is the solution for the free particle problem, the corresponding action

$$S = \int_{t_1}^{t_2} \frac{1}{2}mv^2 dt \tag{1.27}$$

will be extremal, but only for $m > 0$ it will be a minimum, since from this value, any variation $\delta x(t)$ or, consequently, $\delta v(t)$ would make S to increase. Were m negative, S couldn't have a minimum and we would have, instead, a principle of "maximal action". This means that the choice of S having a minimum is connected with the mass m being positive, which is, after all, a nice convention. Anyway, to obtain Lagrange's equations of motion, none of these considerations are relevant, just the extremal condition is enough.

Pursuing further, let us consider the particle interacting with a source of a conservative force. Now, space is no longer homogeneous and the Lagrangian must depend on x, the distance of the particle to the source of the interaction, an scalar.

---

[3] Recall that this is an experimental law, so we are not violating the paradigm that Hamilton's principle gives us all theoretical support.

How can it enter the Lagrangian? Noting that for a free particle L is its kinetic energy, an immediate guess is that the potential energy $V(x)$ be added to it. It is left as an exercise for the readers to test some combinations:

Exercise Test the combinations $T + V$, $T - V$, $T - aV$ and others at your choice, taking them to Lagrange's equation and checking which one works properly.

The reader must have concluded that the appropriate choice is $L = T - V$, for a particle in a conservative field. There is no reason for this recipe be different for the 3D case, or even for multiple particles of an also conservative system. Normally this form for L is justified by it leading to Newton's second law, but here I stress that this is an inconvenient limitation, as forward examples will show. The reader must not be bored with the previous kind of reasoning; in fact the theories of physics are constructed, in general, by an inductive process like this.

Example 1.1 Let us first take the case of a 1D harmonic oscillator and make

$$L = T - V = \frac{1}{2}m\dot{x}^2 - \frac{1}{2}kx^2. \tag{1.28}$$

Using Lagrange's equation we easily obtain,

$$\frac{\partial L}{\partial x} = -kx \quad e \quad \frac{d}{dt}\frac{\partial L}{\partial \dot{x}} = \frac{d}{dt}m\dot{x} = m\ddot{x}, \tag{1.29}$$

so that, with this choice of the Lagrangian, we arrive at the well-known equation of motion of the oscillator

$$m\ddot{x} + kx = 0, \tag{1.30}$$

which is normally obtained from Newton's second law, $F = ma$.

The Lagrangian L of the physical system plays a key role in its dynamics. Since the variational principle that we use invariably leads to the Lagrange equation (1.21), we see that the proper choice of the function $f = L$ becomes vital for the correct description of a mechanical problem.

1.5 "Magic" Lagrangians (One Particle)

The unusual term "magic" used here serves to call the readers attention for the remarkable fact that the form of the Lagrangian is, in principle, to be guessed from symmetry properties, as we will show in some examples. It is never too much to recall that the form $T - V$ contains the conditions demanded by (Galilean) space symmetries, as well as contains $x(t)$ and $\dot{x}(t)$, that is, the variables we need to have our particle mechanical problems solved. What about an explicit dependence of L on t? It will exist in some cases, namely when a particle is subject to a variable field or to a moving constraint (next chapter). But whenever we treat conservative systems, we hopefully have already developed the intuition that an explicit dependence of L on t would not be appropriate.

Example 1.2 Variable mass systems are automatically included in the prescription $L = T - V$. In fact, considering the fall of a drop of water in the gravitational field, with $m_0 = 0$ and $v_0 = 0$ [2], consider the Lagrangian

$$L = \frac{m}{2}\dot{z}^2 + mgz, \tag{1.31}$$

where z is oriented vertically up and m varies on time, in some way to be prescribed. So,

$$\frac{\partial L}{\partial z} = mg, \quad \frac{d}{dt}\frac{\partial L}{\partial \dot{z}} = \frac{d}{dt}(m\dot{z}) = m\ddot{z} + \dot{m}\dot{z}, \quad \Rightarrow \quad m\ddot{z} + \dot{m}\dot{z} = mg, \tag{1.32}$$

which is the well-known equation of motion for a variable mass system.

Example 1.3 In order to challenge, for the first time, the "normal" procedure of taking $L = T - V$, let us, for example, consider a form of L that is explicitly dependent on time but still follows the symmetry prescriptions, which is

$$L = e^{\gamma t}(\frac{1}{2}m\dot{x}^2 - \frac{1}{2}kx^2). \tag{1.33}$$

Although we can directly use the Euler-Lagrange equation for obtaining the equation of motion, it is instructive to show that it can be obtained directly by variation of the action S, even in this case where $L \neq T - V$. We have

$$\delta S[x, \dot{x}, t] = \int_{t_t}^{t_2} [\frac{\partial L}{\partial x}\delta x + \frac{\partial L}{\partial \dot{x}}\frac{d}{dt}\delta x + \frac{\partial L}{\partial t}\delta t]dt = \int_{t_t}^{t_2} [\frac{\partial L}{\partial x} - \frac{d}{dt}\frac{\partial L}{\partial \dot{x}}]\delta x dt =$$

$$\tag{1.34}$$

$$\int_{t_t}^{t_2} [-e^{\gamma t}kx - \frac{d}{dt}(e^{\gamma t}m\dot{x})]\delta x dt = \int_{t_t}^{t_2} [-e^{\gamma t}(kx + \gamma m\dot{x} + m\ddot{x})]\delta x dt = 0, \tag{1.35}$$

which produces the equation of motion,

$$m\ddot{x} + \gamma m\dot{x} + kx = 0. \tag{1.36}$$

In the second passage, we made $\delta t = 0$ (remembering that the variation of x is made for fixed t), used again integration by parts and resorted to the fact that $\delta x(t_1) = \delta x(t_2) = 0$. The resulting Eq. (1.36) is the well-known equation of motion of a harmonic oscillator subjected to a dissipative force of the type $-\gamma m\dot{x}$. The multiplicative term $e^{\gamma t}$ in the Lagrangian led to the representation of a system whose energy is not conserved, as we suspected earlier.

Exercise Try to identify the movement corresponding to $L = e^{2\gamma x}(\frac{1}{2}m\dot{x}^2)$.

At this point, if the reader is amazed that a Lagrangian $L \neq T - V$ led to the proper equation of motion for a known physical system, we must recall that the form $L = T - V$ was by no means a "logical" choice. It was a choice among many others possible. We know from Newtonian mechanics that the equation of motion of a particle $m\dfrac{d^2x}{dt^2} = F$ has been determined experimentally, as well as the forms of external interactions represented by F also have. They didn't come by pure thinking! There is therefore no reason to expect that, in terms of Hamilton's principle, things would be different. We can think that nature chose $L = T - V$ in the first example and $L = e^{\gamma t}(T - V)$ in the second. This is as much as we can say about the previous determination of the Lagrangian, based on our knowledge of what we call *symmetries* of the system and literally guessing a form for L. The student must abandon the idea that $L = T - V$ is a logical form while the others would be mere curiosities. To illustrate further this fundamental point, as well as to widen its scope, we develop one more example and, after that, introduce the concept of generalized potentials.

Example 1.4 Let us now take the case of a particle subject to a conservative potential $V(r)$, but in the relativistic regime. The "trial" Lagrangian

$$L = -mc^2(1 - \frac{v^2}{c^2})^{\frac{1}{2}} - V(r) \tag{1.37}$$

is evidently different from $T - V$, but should reduce to this form for $c \to \infty$, since it must reproduce the non-relativistic Lagrangian. In fact, the series expansion of the first term, taking into account that $c >> v$ leads, to first-order, to

$$L \approx -mc^2 + \frac{1}{2}mv^2 - V(r), \tag{1.38}$$

where the first term is constant, having thus no effect on the equation of motion or on Hamilton's principle. In turn, from Lagrangian (1.37) we get the equation of motion

$$\frac{d}{dt}\frac{m\vec{v}}{\sqrt{1 - \frac{v^2}{c^2}}} = -\vec{\nabla}V(r), \tag{1.39}$$

which is the correct non-covariant equation of motion in the relativistic regime and which is reduced to Newton's second law equation for $c >> v$. We can see this result, together with the previous examples, as indicating that the conservative Lagrangian $L = T - V$ should be a special case of more general Lagrangians.

Exercise Using components x, y, z, deduce Eq. (1.39) from Lagrangian (1.37).

Let us go deeper in analysing Lagrangian (1.37). We know from special relativity that the proper time τ is related to t by

$$d\tau = \sqrt{1 - \frac{v^2}{c^2}}dt. \tag{1.40}$$

So, in view of Eq. (1.37), the corresponding action can be written as

$$S = -mc^2 \int d\tau - \int V(r)dt. \tag{1.41}$$

Paying attention just to the first term for now, what can we get from it? τ is a Lorentz invariant quantity as we know, and we expect S to also be. It means that some of the "magic" in guessing Lagrangian (1.37) has now been revealed: S must be Lorentz invariant and τ is an invariant quantity at hand. One more time, a symmetry property induces us to choose a proper Lagrangian, so that Eq. (1.37) is not guessed out by luck, but rather the result of a reasonably oriented inductive process.

But, what about V, is any known potential $V(r)$ acceptable? The answer is no!, S must be Lorentz invariant and static potentials are not. This feature reveals itself clearly in case of the electrostatic potential and this should not be a surprise. Electrostatic interaction is assumed to be instantaneous. Moreover, the Coulomb potential is valid only in a reference frame in rest relative to the source charges, not for all inertial frames where charges are moving. This matter is treated properly in terms of fields. Before we do this, it is convenient to change our notation.

1.6 Adapting Notation

Just for notational convenience for the moment, from this point on we will make the identification

$$q_k(t) \equiv x_{ki}(t), \quad k = 1, 2, 3 \ldots, n \quad i = x, y, z. \tag{1.42}$$

Except in the presence of constraints to the movement of the system (see next chapter), q_k will be representing the entire set of vector coordinates, sometimes represented simply as q, even in the case of systems with more than one particle. In general, the symbol q will represent both a set of coordinates or a single one, depending on the context. Then, consider a system of particles, all of them free to move in 3D (or lower dimensions, 1D, 2D). We want to apply Hamilton's principle to predict their movements. The Lagrangian will be $L = L(q, \dot{q}, t)$, so that variation of the action yields,

$$\delta S = \int_{t_t}^{t_2} \delta L(q, \dot{q}, t) dt = \int_{t_t}^{t_2} \sum_{k=1}^{n} [\frac{\partial L}{\partial q_k} \delta q_k + \frac{\partial L}{\partial \dot{q}_k} \delta \dot{q}_k] dt = 0. \tag{1.43}$$

As we evaluate the integral of the second term by parts, in a similar way as we did previously, Hamilton's principle yields

$$\delta S = \int_{t_t}^{t_2} \delta L(q, \dot{q}, t) dt = \int_{t_t}^{t_2} \sum_{k=1}^{n} [\frac{\partial L}{\partial q_k} + \frac{d}{dt} \frac{\partial L}{\partial \dot{q}_k}] \delta q_k dt = 0. \tag{1.44}$$

Now, since all q_k are independent of each other, all δq_k are independent too, and are also arbitrary. The only way Eq. (1.44) can be zero is each quantity in brackets being zero, independently. We conclude by the existence of one Lagrange's equation for each coordinate q_k, that is,

$$\frac{\partial L}{\partial q_k} - \frac{d}{dt} \frac{\partial L}{\partial \dot{q}_k} = 0. \quad k = 1, 2, 3 \ldots n \tag{1.45}$$

This set of equations is valid for all coordinates of particles in 3D (or lower) motion. In the next chapter we will treat problems with constraints and define the *generalized coordinates* {q} in a more restricted way.

Exercise Check that the Lagrangian

$$L = T - V = \frac{1}{2}m(\dot{x} + \dot{y}^2) - \frac{1}{2}k(x^2 + y^2) \tag{1.46}$$

produce the 2D isotropic harmonic oscillator equations of motion.

1.7   Generalized Potentials

Back to the discussion about potentials, we saw that static potentials $V(r)$ cannot be Lorentz invariant. The fact that this drawback is overcome by a generalization of the $L = T - V$ recipe is another spectacular example of how physics works. The case of a particle in an electromagnetic field, illustrates further this point by generalizing the potential, rather than the kinetic energy term. Two important symmetry properties of the Lagrangian will be introduced. Let us do it formally first.

In Cartesian coordinates, T depends only on velocities $\dot{q}_k$ (after introducing the generalized coordinates, this will be changed), while V depends only on the coordinates q_k. Lagrange's equations (1.45), can thus be written as

$$\frac{d}{dt}\frac{\partial T}{\partial \dot{q}_k} = -\frac{\partial V}{\partial q_k} \equiv Q_k, \tag{1.47}$$

whose right-hand side represents the components Q_k of the applied force $\vec{Q}$. Let us consider, on the other hand, that the vector force is derived from a velocity-dependent generalized potential $U(q, \dot{q})$, in a way that its components are

$$Q_k = -\frac{\partial U}{\partial q_k} + \frac{d}{dt}\frac{\partial U}{\partial \dot{q}_k}. \tag{1.48}$$

By extention, since T does not depend on the q_k, we can consider the equation of motion as being

$$\frac{d}{dt}\frac{\partial T}{\partial \dot{q}_k} - \frac{\partial T}{\partial q_k} = -\frac{\partial U}{\partial q_k} + \frac{d}{dt}\frac{\partial U}{\partial \dot{q}_k}, \tag{1.49}$$

or,

$$\frac{\partial (T-U)}{\partial q_k} - \frac{d}{dt}\frac{\partial(T-U)}{\partial \dot{q}_k} = 0. \tag{1.50}$$

Thus, if the Lagrangian is now defined as $L = T - U$, the equation above is such that the motion of the particle follows from Hamilton's principle.

Example 1.5 The apparently formal generalized potential proposal, has its correspondence "simply" for the motion of a charged particle in an electromagnetic field. Moreover, the procedure is fundamental to account for the relativistic nature of this interaction. To show this consider the Lagrangian as

$$L = T - U = \frac{m}{2}v^2 - q\Phi + \frac{e}{c}\vec{v}.\vec{A}, \tag{1.51}$$

where e is the particle's electric charge, c the speed of light and $\Phi(\vec{r}, t)$ and $\vec{A}(\vec{r}, t)$ are, respectively, the electrostatic scalar potential and the vector potential, from which the fields are derived by,

$$\vec{E} = -\vec{\nabla}\Phi - \frac{1}{c}\frac{\partial \vec{A}}{\partial t}, \quad \vec{B} = \vec{\nabla} \times \vec{A}. \tag{1.52}$$

Introducing the compact notation,

$$\frac{\partial}{\partial \vec{v}} \equiv \frac{\partial}{\partial \dot{x}}\vec{i} + \frac{\partial}{\partial \dot{y}}\vec{j} + \frac{\partial}{\partial \dot{z}}\vec{k}, \tag{1.53}$$

which can be seen as the inverse operation of the scalar product of a generic vector by $\vec{v}$ (check it!), we get $\dfrac{\partial L}{\partial \vec{v}} = m\vec{v} + \dfrac{q}{c}\vec{A}$, and $\dfrac{\partial L}{\partial \vec{r}} = \vec{\nabla} L = -e\vec{\nabla}\Phi + \dfrac{e}{c}\vec{\nabla}(\vec{v}.\vec{A})$.

Using the identity,

$$\vec{\nabla}(\vec{v}.\vec{A}) = \vec{v}\times(\vec{\nabla}\times\vec{A}) + (\vec{v}\cdot\vec{\nabla})\vec{A}, \tag{1.54}$$

we show, after some manipulation, that the vector Lagrange equation becomes

$$\frac{d}{dt}(m\vec{v} + \frac{e}{c}\vec{A}) = -q\vec{\nabla}\Phi + \frac{e}{c}\left[\vec{v}\times(\vec{\nabla}\times\vec{A}) + (\vec{v}\cdot\vec{\nabla})\vec{A}\right]. \tag{1.55}$$

This is the equation of motion for charge q in an electromagnetic field in the Lagrangian theory. We can put it in the known Newtonian form, verifying that the total time derivative of vector $\vec{A}$ can be written as,

$$\frac{d\vec{A}}{dt} = (\frac{\partial\vec{A}}{\partial x}\dot{x} + \frac{\partial\vec{A}}{\partial y}\dot{y} + \frac{\partial\vec{A}}{\partial z}\dot{z}) + \frac{\partial\vec{A}}{\partial t} = (\vec{v}\cdot\vec{\nabla})\vec{A} + \frac{\partial\vec{A}}{\partial t}. \tag{1.56}$$

Taking this result to the left-hand-side of Eq. (1.55) we can write,

$$\frac{d}{dt}(m\vec{v}) = -e(\vec{\nabla}\Phi + \frac{1}{c}\frac{\partial\vec{A}}{\partial t}) + \frac{e}{c}\vec{v}\times(\vec{\nabla}\times\vec{A}), \tag{1.57}$$

where the terms $\pm\dfrac{e}{c}(\vec{v}\cdot\vec{\nabla})\vec{A}$ were cancelled, finally getting, in view of Eqs. (1.52) above, to

$$m\frac{d\vec{v}}{dt} = e(\vec{E} + \frac{\vec{v}}{c}\times\vec{B}), \tag{1.58}$$

which is Newton's equation for the charged particle under the action of the Lorentz force.

It is evident that the form of the generalized potential in Eq. (1.51) was chosen for this purpose, but the fact that the description of a fundamental interaction of nature coincides with a formal generalization of the standard Lagrangian to include a velocity dependent term is a remarkable demonstration of the power of Lagrangian mechanics.

But something seems still awkward: We cannot use the electrostatic potential $V(\vec{r}) \equiv \Phi(\vec{r})$ in the relativistic Lagrangian because it is not Lorentz invariant, but using the generalized relativistic potential in the non-relativistic Lagrangian we recover Newton's equation! How can it be? The answer is simple, however: Lorentz transformations becomes Galilean transformations in the limit $v \ll c$ (note the c^{-1}

in the last term of Eq. (1.51)) but the opposite is not true. It is a matter of *status*! On the other hand, in the relativistic regime the "kinetic" term of the Lagrangian must also change, as follows.

Example 1.6 For the charged particle in an electromagnetic field in the relativistic regime, we substitute the kinetic energy of the previous Lagrangian by the total energy of the relativistic free-particle,

$$L = -mc^2(1 - \frac{v^2}{c^2})^{\frac{1}{2}} - q\Phi + \frac{q}{c}\vec{v} \cdot \vec{A}. \tag{1.59}$$

Lagrange's equation of motion, in vector form, is obtained as,

$$\frac{d}{dt}\left[\frac{m\vec{v}}{\sqrt{1 - \frac{v^2}{c^2}}} + \frac{q}{c}\vec{A}\right] = \vec{\nabla}(-q\Phi + \frac{q}{c}\vec{v} \cdot \vec{A}), \tag{1.60}$$

which, by a completely analogous procedure to the previous one, can be written,

$$\frac{d}{dt}\left[\frac{m\vec{v}}{\sqrt{1 - \frac{v^2}{c^2}}}\right] = q(\vec{E} + \frac{\vec{v}}{c} \times \vec{B}). \tag{1.61}$$

Voilá, this is the correct non-covariant relativistic equation of motion for a point charge in an electromagnetic field.

Exercise Derive Eqs. (1.60) and (1.61) in detail.

The reader should be now aware of a common feature of the apparently disconnected examples presented: Once specific Lagrangians for each problem are found, generating the correct solution through Lagrange's equations, these equations of motion follow as consequences of the a single common source, Hamilton's principle, which is then considered the universal element behind all classical motions.

A paradigmatic view of what has been exposed so far is that, once the validity of Hamilton's principle is assumed, *there must be a Lagrangian* that generates the correct dynamics of the studied system. The corollary is that, once the correct Lagrangian is found, Hamilton's principle guarantees obtaining the proper equation of motion.

This point of view lends the Lagrangian an heuristic but fundamental character. There seems to be no doubt that this paradigm is vastly more appropriate than to assume that the Lagrangian is always $L = T - V$ and the action principle is restricted to conservative non-relativistic systems, the cases outside being simply coincidences.

1.8 Equivalent Lagrangians

We learned that in the investigation of the possible forms of Lagrangians of specific systems, their symmetries play a central role. Formally, this discussion will be made in the scope of Nöther's theorem, in the following chapter, but there is another important property to be advanced, which leads us to the concept of *equivalent Lagrangians*. As we know, the kinetic energy of a non-relativistic particle is not invariant under a Galilean velocity transformation. However, the equations of motion are necessarily invariant under this transformation, so, the resulting action S must be invariant. Hence, we expect that the action integral displays a property connected to this symmetry.

If a Lagrangian L and another L' differ by a total time derivative of a function of coordinates and time $\dfrac{d}{dt}g(q,t)$, that is,

$$L' = L + \frac{d}{dt}g(q,t), \tag{1.62}$$

they will generate the same equations of motion, In fact,

$$\delta S' = \delta \int_{t_t}^{t_2} L(q,\dot{q},t)dt + \delta \int_{t_t}^{t_2} dg(q,t) = \delta S, \tag{1.63}$$

since $q = q(t)$ and the variation of the second term, at fixed extremes, vanishes.

We can carry out the same proof directly from Lagrange equations. For a generic q coordinate,

$$\frac{d}{dt}\frac{\partial L'}{\partial \dot{q}} = \frac{d}{dt}\frac{\partial L}{\partial \dot{q}} + \frac{d}{dt}\frac{\partial}{\partial \dot{q}}\frac{d}{dt}g = \frac{d}{dt}\frac{\partial L}{\partial \dot{q}} + \frac{d}{dt}\frac{\partial}{\partial \dot{q}}(\frac{\partial g}{\partial q}\dot{q} + \frac{\partial g}{\partial t}) = \tag{1.64}$$

$$\frac{d}{dt}\frac{\partial L}{\partial \dot{q}} + \frac{d}{dt}\frac{\partial g}{\partial q}.$$

On the other hand,

$$\frac{\partial L'}{\partial q} = \frac{\partial L}{\partial q} + \frac{\partial}{\partial q}\frac{dg}{dt} = \frac{\partial L}{\partial q} + \frac{d}{dt}\frac{\partial g}{\partial q}, \tag{1.65}$$

since the operators $\dfrac{d}{dt}$ and $\dfrac{\partial}{\partial q}$ commute. Hence, it follows that if the Lagrange equation holds for L, it also applies to L'.

Example 1.7 Let us see an application of paramount importance in classical theory of fields. We saw in Example 1.5 that the fields $\vec{E}$ and $\vec{B}$ derive, respectively, from the potentials Φ and $\vec{A}$ by Eqs. (1.52). It is straightforward to verify that the transformed potentials

$$\Phi' = \Phi - \frac{1}{c}\frac{\partial \Omega(\vec{r},t)}{\partial t}, \quad \vec{A}' = \vec{A} + \vec{\nabla}\Omega(\vec{r},t), \tag{1.66}$$

generate the same fields $\vec{E}$ and $\vec{B}$, for a generic function $\Omega(\vec{r},t)$.

Exercise Check it!

The idea is that the fields, not the potentials, are the actual physically observables (not in the scope of quantum mechanics, however). The potentials are said then to depend on the choice of a *gauge*. Taking these expressions to Eq. (1.51) to produce L', we get

$$L' = L + \frac{q}{c}\left(\frac{\partial \Omega}{\partial t} + \vec{v}\cdot\vec{\nabla}\Omega\right) = L + \frac{q}{c}\frac{d}{dt}\Omega(\vec{r},t), \tag{1.67}$$

which guarantees that the same equations of motion will be produced by L' and L, as expected. Later in the text, we will show another important consequences of this property of the Lagrangian.

Exercise A q_l coordinate that does not appear in the Lagrangian L (called cyclic coordinate, see next chapter) may appear in the equivalent Lagrangian $L' = L + \dfrac{d}{dt}f(q,t)$. Show that the equations of motion will still be the same.

1.9 Newton's Laws from Hamilton's Principle

While the equivalence of Lagrange's equations and Newton's second law catches the eyes (and will be formally proved in the next chapter), the curious student may well ask how the other two laws, also fundamental, are derived from Hamilton's principle, which they *must* do, in view of what we have seen so far. This point is discussed in what follows.

1.9.1   Newton's First Law: An Equivalent Lagrangian

Consider a particle of mass m whose 1D motion is followed by two observers moving in the same direction, with their uniform velocities differing by u. The velocities of the particle for the two observers are related by $v' = v + u$ (Galilean transformation). Since an eventual potential energy V of the particle does not differ for the two observers, they will obtain their respective Lagrangian functions differing only by their kinetic energies,

$$L' = \frac{1}{2}mv'^2 - V = \frac{1}{2}m(v+u)^2 - V = \frac{1}{2}mv^2 + muv + \frac{1}{2}mu^2 - V$$

$$= L + 2vu + C, \tag{1.68}$$

where C is a constant. In consequence, the variation of the action integrals for L' and L are the same, in view of, for the term $2vu$,

$$\delta S = \delta \int_{t_1}^{t_2} 2u \frac{dx}{dt}\, dt = 2u\delta[x(t_2) - x(t_1)] = 0, \tag{1.69}$$

since Hamilton's principle demands the extreme points of the paths, at t_1 and t_2, to be fixed.

That is, the two Lagrangian functions are equivalent and, so, generate the same equation of motion. The conclusion is that the two observers cannot detect any difference in the equations of motion of the particle. In other words, their motions are not absolute, they stand as inertial observers. Despite the difference of terminology in Newton's statement of its first law, it is consensual today that this law corresponds to establishing of inertial frames.

Just checking, Lagrange's equation applied to L' yield $\dfrac{\partial L'}{\partial x} = \dfrac{\partial L}{\partial x}$ and $\dfrac{d}{dt}\dfrac{\partial L'}{\partial v} = \dfrac{d}{dt}\dfrac{\partial L}{\partial v}$, producing obviously the same equations of motion.

1.9.2   Newton's Third Law: Space Homogeneity

Let us consider an isolated system of two interacting particles with masses m_1 and m_2. Defining the center-of-mass $\vec{R}$ and relative $\vec{r}$ vectors, and the corresponding velocities $\vec{V}$ and $\vec{v}$ as usual, it is a common exercise to obtain the Lagrangian as

$$L = \frac{1}{2}M\vec{\dot{R}}^2 + \frac{1}{2}\mu\vec{\dot{r}}^2 - V(r) \tag{1.70}$$

where $M = m_1 + m_2$ is the total mass, $\mu = \dfrac{m_1 m_2}{m_1 + m_2}$ the reduced mass and the potential energy V depends just on the distance r of the particles.

Exercise Prove it.

Observing that since any component of $\vec{R}$, say X, is cyclic (does not appear explicitly in L), it follows from Lagrange's equations that $\dfrac{d}{dt}\dfrac{\partial L}{\partial \dot{X}} = 0$, meaning that $P_X = \dfrac{\partial L}{\partial \dot{X}}$ is constant, so does $\vec{P} = M\vec{V}$. But $\vec{P} = \vec{p_1} + \vec{p}_2$, which yields $\dot{\vec{p}}_1 = -\dot{\vec{p}}_2$.

In view of the definition of force, $\vec{F} = \dot{\vec{p}}$, Newton's third law is obtained, $\vec{F}_{12} = -\vec{F}_{21}$, where $\vec{F}_{ij}$ means the force caused by particle j on particle i. Newton's third law is then a consequence of space homogeneity, or, equivalently, of the conservation of the total linear moment, which is accounted for in Lagrangian mechanics (that is, from Hamilton's principle) by the cyclic character of the center-of-mass coordinates.

1.10 Connection to Quantum Mechanics: The Classical Limit

Extending Hamilton's principle to quantum mechanics seems out of question at a first sight, since this principle works by choosing the best trajectory amongst an infinite number of other possible between two instants, but in quantum mechanics there is no such thing as "best trajectory". In fact, there is no trajectory at all, at least in the classical sense. Our source of information about the movement of a quantum system is a complex *wavefunction* $\psi(\vec{r}, t)$, from which we interpret the *probability density* for the system as $\psi^* \psi$. This means that the volume integral of this density gives the probability of finding the system inside that volume.

In turn, Feynmann [3] considered that the infinite trajectories of classical mechanics, should be kept in quantum mechanics. In this sense there is no similarity between the two theories, on the contrary. But, at the same time, he postulated a very interesting connection between them in which the action S plays, again, a central role.

Let us introduce the quantum *propagator* $K(\vec{r}\,', t'; \vec{r}, t)$, that controls the temporal evolution of a quantum particle so that

$$\psi(\vec{r}\,', t') = \int K(\vec{r}\,', t'; \vec{r}, t)\psi(\vec{r}, t)\, d^3\vec{r}. \tag{1.71}$$

Feynmann postulated a connection of the propagator with the action S for all trajectories as

$$K(\vec{r}\,', t'; \vec{r}\,', t) = \sum_{\alpha} exp(\frac{i}{\hbar} S_{\alpha}) \qquad (1.72)$$

where the continuous index α ranges over all possible trajectories connecting points 1 and 2.

Consider that all values S_{α} exceeds $\hbar$ by many orders of magnitude. The presence of the imaginary unit i in Eq. (1.72) implies oscillating terms that will interfere destructively except in the vicinity of the trajectory correspond to minimum S. In the classical limit, $\hbar$ goes to zero and the trajectories reduce to a single one, that one correspondent to the minimal S of Hamilton's principle. So, in spite Hamilton's principle itself not applying to the microscopic world, the action S still plays a role in the dynamics of quantum systems. Moreover, the possibly disappointed reader may find solace in the fact that the time-independent Schrödinger equation is obtained from a variational principle: The wavefunctions of the stationary states (the "orbits" of the electron in an H atom, for instance) are those that minimize their energies. This variational property is quite useful in quantum physics.

1.11   Final Considerations on General Lagrangian Mechanics

The "field" delimited by $L = T - V$, meaning the mechanics of classical conservative systems, has its rightful place, but then, equivalence of the Newtonian and Lagrangian theories must be proved fully, for the three Newton's laws, not just for the second law. Once this is done, the two theories show equivalent outcomes. There are, of course, alternative views to this point. Starting from conservation principles the situation is reversed. For example, the third law does not appear as an independent law of nature but a particular result of the principle of conservation of the linear momentum for a two-particle system, related to the symmetry property of space homogeneity. What is remarkable is that all this is contained in Hamilton's principle.

Solving problems directly by the variational method is not necessarily a more operational procedure, inasmuch as that, in general, it leads to differential equations that must be solved anyway. What it does, however, is extraordinary: It tells us that, once the equation of motion is solved, the solution $\vec{r}(t)$ is the one that minimizes the functional J. As the same principle leads to different mechanics, namely Newtonian, relativistic and field theory (in Chap. 4), it is immediate to think of it as having a superior *status* as compared to particular laws of motion, which are deduced from it. The combination of Hamilton's variational principle with the appropriate choice of the Lagrangian demonstrates, then, to be a powerful theoretical tool, which expands the limits of classical Newtonian mechanics, making use of the symmetries of nature.

A further remarkable example of the superior *status* of Hamilton's principle, namely, working properly in non-inertial reference frames, is postponed to the next chapter, since it has close connection to the presence of (accelerated) constraints, thus demanding the introduction of *generalized coordinates*.

Chapter 2
Lagrangian Mechanics of Systems with $L = T - V$

Abstract Moving to constrained systems with many degrees of freedom, standard Lagrangian theory is addressed, at a level appropriate to teach even at the undergraduate level. However, the particular focus on underlying topics of reasonable difficulty for students is kept for two issues: (i) Moving constraints and accelerated reference frames, and (ii) Galilean invariance, for which derivations normally absent in standard texts, or even a correction of a common misconception, are presented.

Keywords Constraints · Generalized coordinates · Lagrange's equations · Accelerated constraints · Nöther's theorem · Galilean invariance of Jacobi's integral · Rigid bodies motion

In Lagrangian mechanics, special emphasis is due to systems of N particles, for which the Lagrangian can always be written as

$$L = T - V, \tag{2.1}$$

where V is the total scalar potential energy (or simply the *potential*),

$$V = \sum_{\alpha=1}^{N} V_\alpha, \quad \vec{F}_\alpha = -\vec{\nabla} V_\alpha, \tag{2.2}$$

being $\vec{F}_\alpha$ the forces that act on the individual particles, represented by the index α. Evidently, conservative systems form a class of problems of special interest in this chapter.

On the other hand, an explicit dependence of L on time can come also from the presence of moving constraints to the movement of the particles, which will be reflected in their kinetic energy T. In this case, the subject of energy conservation will become more complex, as we shall see.

In order to discuss problems with multiple degrees of freedom and that involve unknown constraint forces, it is necessary to introduce some characteristic concepts of Lagrangian mechanics.

2.1 Constraints and Generalized Coordinates

We have already seen that it is impossible to reduce all mechanical systems, even approximately, to a model of classical punctual particles interacting with each other. There are geometric or kinematic *constraints* to the movements of the particles. These, when can be described by equations involving only particle coordinates and, eventually, time, are the so-called *holonomic* constraints which, in general, represent geometric obstacles to particle's motion.

They exert forces on the particles, the *constraint forces*, which are previously unknown, so that problems with multiple particles, under the action of constraints, can become extremely complicated in Newtonian mechanics.

An important warning is that dissipative forces are not constraint forces, since they act dynamically on the systems subject to them. Empirical formulas are need for them, for example, the velocity dependent friction forces.

The attempt to eliminate unknown holonomic constraint forces from Newton's equations, due especially to D'Alembert, led to the first developments of analytical mechanics. To motivate this question, let us consider the problem of a particle moving under the action of a conservative force $\vec{F}$ and, also, of a non-dissipative constraint force $\vec{R}$, that is, such that $\vec{R} \cdot \vec{v} = 0$. More specifically, let us consider that the constraint compels the particle to a path described by a known curve (a bead sliding on a wire, as in Fig. 2.1, for example).

From Newton's second law,

$$\vec{F} + \vec{R} = m\frac{d\vec{v}}{dt} \quad \Rightarrow \quad \vec{F} \cdot \vec{v} = m\frac{d\vec{v}}{dt} \cdot \vec{v} \quad \Rightarrow \quad -\vec{\nabla} V \cdot d\vec{r} = d(\frac{1}{2}mv^2),$$

$$(2.3)$$

where $V(r)$ is the potential energy associated to the force $\vec{F}$. Integrating this equation, we obtain

$$\frac{1}{2}mv^2 + V(x, y, z) = E, \tag{2.4}$$

where the constant E is the mechanical energy, dependent only on the initial conditions. On the other hand, the given the parametric equations of the curve in terms of distance s measured over it,

$$x = x(s), \quad y = y(s), \quad z = z(s), \tag{2.5}$$

and, since $ds = \sqrt{dx^2 + dy^2 + dz^2}$, we are lead to

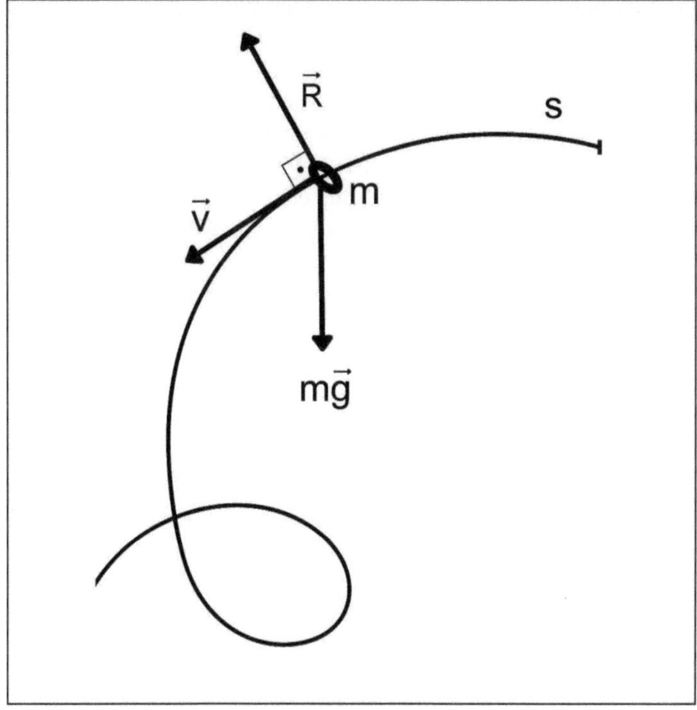

Fig. 2.1 Non-dissipative constraint force $\vec{R}$ on a bead in a wire

$$\frac{1}{2}m(\frac{ds}{dt})^2 + V(s) = E. \tag{2.6}$$

By this procedure, we have a scalar equation, from which we can obtain $v(s)$ and then, $v(t)$ and $s(t)$. Hence, as a result of eliminating the constraint force from the equation of motion, we reduce the problem from 3D to a single curvilinear dimension.[1] This feature becomes even more illustrative when we differentiate this equation with respect to s, which leads to Newton's 1D equation,

$$m\frac{d^2s}{dt^2} = F(s). \tag{2.7}$$

This simple procedure, of eliminating the orthogonal constraint force in the case of a single particle, was generalized to more complex mechanical systems by D'Alembert, as will be discussed further on. For the time being, it should be noted that the term "elimination" of constraint forces refers only to their absence in the

---

[1] For this, the constraint force must be 2-dimensional.

equations of motion; evidently these forces keep their action on the motion of the system, restricting it in different ways.

Finally, it should be noted, in the previous example, that the coordinate s is not a component of the particle's position vector, and the equation of motion is also not vector. This is a characteristic of Lagrangian mechanics, it is a *scalar mechanics*. In order to develop it to multi-particle systems, with multiple curvilinear coordinates, we need to introduce some further important concepts.

For simplicity of language, in this book we refer equally to the constraints themselves or to the bodies that impose them, under the same term, *constraints*, even being aware that this nomenclature is not the most elegant.

2.1.1 Holonomic Constraints

As already said, constraint forces are those unknown forces, generally from a large lot of microscopic electromagnetic interactions, which constrains the movement of particles in some way, but are not the driving forces responsible for the dynamic motion. The so-called normal reaction, tension and static friction are examples of constraint forces. In N-particle multi-dimensional mechanical systems, holonomic constraints, in number p, are those described by equations like

$$g_l(x_1, \ldots, x_{3N}, t) = 0, \quad l = 1, \ldots, p, \tag{2.8}$$

that is, those involving only the coordinates of the particles and, eventually, time (rheonomic holonomic constraints). One aspect that it is never too much to clarify is that the coordinates x_i involved in holonomic constraint equations are not necessarily cartesian, but necessarily determine all particle's position vectors, that is, they exist in number 3N.

Constraints described by inequalities or by velocity dependent differential equations, non-integrable to a form like (2.8), are called *non-holonomic* and require special theoretical handling. They restrict possible displacements, but not necessarily the possible system configurations.

There are cases, though less common, of constraints involving particle velocities, but which can be reduced by integration to the holonomic form (2.8). Clearly, if a constraint equation has the form

$$dg = \sum_j \frac{\partial g}{\partial x_j} \dot{x}_j + \frac{\partial g}{\partial t} = 0, \tag{2.9}$$

then, immediate integration leads to

$$g(x_j, t) + C = 0, \tag{2.10}$$

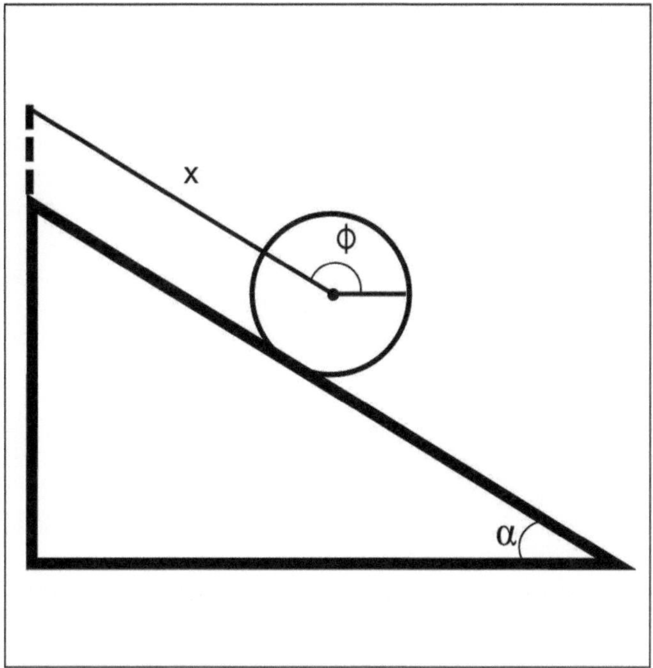

Fig. 2.2 Disk rolling down an inclined plane

which is an equation of a holonomic constraint. A classic case is that of a cylinder rolling without slipping on an inclined plane, shown in Fig. 2.2,

Integrating the constraint equation,

$$\dot{x} - R\dot{\phi} = 0 \tag{2.11}$$

we obtain,

$$g(x, \phi) = x - R\phi = 0, \tag{2.12}$$

where the constant of integration is conveniently taken as zero. This equation represents a holonomic constraint between the Cartesian coordinate x of the CM of the cylinder and the angle ϕ of its rotation around its axis.

2.1.2   *Generalized Coordinates an Velocities*

In 3D space, a N-particle system subject to p holonomic constraints has its configurations restricted to its number of *degrees of freedom*, n, in most simple cases determined by

$$n = 3N - p. \tag{2.13}$$

This is the (maximum) number of coordinates needed to specify each configuration, or, equivalently, the dimension of the configuration space (see next subsection). However, this formula must not be taken to the letter. Particularly, it will fail in the presence of rigid bodies.

Consider, for example, a model rigid-body composed of N rigidly bonded particles moving in space. By simple combinatorial analysis, and considering that for each pair of particles there is just one constraint, the total number of constraints is $p = \frac{1}{2}C_2^N = \frac{1}{2}\dfrac{N!}{2!(N-2)!}$ so that, using Eq. (2.13) we get $n = \dfrac{13N - N^2}{4}$, which is obviously wrong. The point is that we can count at most three constraints *per* particle, the rest being redundant. As we will see, a rigid-body has $n = 6$ degrees of freedom. What we should keep in mind is that the number of generalized coordinates is equal to n, the number of degrees of freedom of the system, regardless Eq. (2.13) applies or not.

Generalized coordinates are sets of n (not necessarily vector) coordinates *independent* from each other,

$$\{q_k\} \equiv q_k, \quad k = 1, \ldots, n, \tag{2.14}$$

which specify a system *configuration*, or a point of the n-dimensional *configuration space*, see below.

The independence among the generalized coordinates is granted by the inexistence of holonomic constraints between them. As each holonomic constraint equation type (2.8) can be used to "eliminate" a Cartesian (or vector) coordinate, the generalized coordinates are not, in general, components of the position vector. It should also be realized that there is not a unique set of generalized coordinates. However, some sets will be special, as will be seen ahead.[2] Later we will see that symmetries of mechanical problems can also lead to a decrease in the dimensionality of the configuration space, through the emergence of redundant, though still independent, coordinates.

The rates of change of the generalized coordinates in time,

$$\dot{q}_k = \frac{dq_k}{dt}, \tag{2.15}$$

---

[2] In the same way we sometimes use 3N coordinates x_i to represent the N (x, y, z) coordinates of all particles, now we will use n generalized coordinates q_k for them, so that the sum over particle index α will seldom appear.

are called *generalized velocities*. Generalized coordinates are not necessarily distances, they can be also angles, so generalized velocities can be either linear or angular velocities.

Clearly, in the absence of constraints, $n = 3N$, and the generalized coordinates will be the common vector coordinates, Cartesian for instance, as we used in the first chapter.

2.1.3 Configuration Space

The n-dimensional space, containing an axis for each generalized coordinate, is called *configuration space*. The sequence of points in configuration space represents a trajectory of the system in this space. As the system evolves in time, these points change positions, setting a trajectory in the configuration space, which normally is not the trajectory that we see in normal space. Moreover, in systems of many particles, each one draws its own physical trajectory, but the trajectory in configuration space is unique, multidimensional. An example is the plane double-pendulum, for which the configuration space is given by two angles $\theta_1(t)$, $\theta_2(t)$, while the physical trajectory of the pendulum bubbles is given by $x_1(t)$, $y_1(t)$, $x_2(t)$, $y_2(t)$. We can refer also to the parametric trajectory in configuration space adding an axis for the parameter time (see Fig. 2.4 ahead). This choice is useful for illustrate trajectories variations when using Hamilton's principle.

2.1.4 Transformation Equations

Once the holonomic constraint equations are identified, the number n of generalized coordinates is given by Eq. (2.13). As a set of n generalized coordinates is chosen for a system, we do not use the constraint equations anymore, only must transform the Cartesian (or vector) coordinates to generalized coordinates. The mathematical relations between them,

$$\vec{r}_\alpha = \vec{r}_\alpha(q_1, q_2, \ldots, q_n, t), \quad \alpha = 1, \ldots, N, \tag{2.16}$$

are called *transformation equations*. As a consequence, the velocities are given by

$$\vec{v}_\alpha = \frac{d\vec{r}_\alpha}{dt} = \sum_{k=1}^{n} \frac{\partial \vec{r}_\alpha}{\partial q_k} \dot{q}_k + \frac{\partial \vec{r}_\alpha}{\partial t}. \tag{2.17}$$

This relationship will be useful in later developments.

Constraint equations and transformation equations are obviously connected. For example, consider a simple pendulum with length l, making an angle θ with the vertical ($y = 0$ at the pendulum hinge). The constraint equations are $z = 0$ (because the pendulum is fixed) and $\sqrt{x^2 + y^2} - l^2 = 0$. The number of degrees of freedom is $n = 1$ and θ is normally the chosen generalized coordinate. The transformation equations are $x = l\,sen\theta$ and $y = -l\,cos\theta$. If we eliminate θ from these two equations we recover the constraint equation.

2.2 The Lagrangian

For the same reasons presented in the previous chapter, we assume that the Lagrangian of a system of n degrees of freedom depends on the generalized coordinates qs, the generalized velocities $\dot{q}s$ and, eventually, depends explicitly on time t. For Newtonian "conservative" systems, it turns out that the Lagrangian is given by

$$L = T - V = \frac{1}{2} \sum_{\alpha}^{N} m_\alpha \vec{v}_\alpha^{\,2} - V(r_\alpha, t). \tag{2.18}$$

Different from Example 1.3, here the possible explicit dependence on t refers to a time-dependent potential energy, for example, a charged particle in an oscillating electric field. For lacking of a better term, the term "conservative" here is purely formal, meaning that the corresponding force is given by minus the gradient of this potential but, evidently, the mechanical energy is not conserved, it can be being drained from or filled to the system. This possibility fits perfectly to $L = T - V$ but is not special to the Lagrangian formulation, with or without constraints, since Newtonian mechanics also accounts for time-dependent forces.

From the above definition and using Eqs. (2.16) and (2.17), we easily obtain the Lagrangian as a function of generalized coordinates, generalized velocities and time,

$$L = L(q, \dot{q}, t). \tag{2.19}$$

As time evolves, the system goes through successive configurations $q_k(t)$, which form a trajectory in the configuration space. Once again, the reader should notice that these trajectories are not, normally the actual trajectories in our 3D space.

2.2.1 Virtual Displacements

In the derivation of Lagrange's equations from *D'Alembert's principle of virtual works*, the concept of virtual displacement is vital. In turn, in the application of Hamilton's principle to systems subject to constraints, this concept is no longer

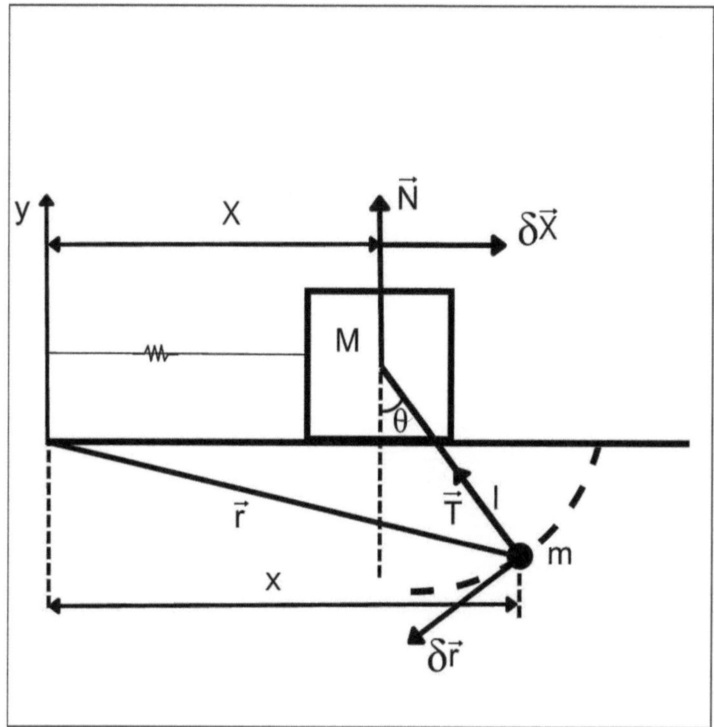

Fig. 2.3 Virtual displacements are perpendicular to the corresponding constraint forces

necessary, but remains illustrative of how constraint forces are ruled out from each problem just by choosing independent generalized coordinates.

Let us consider the $n=2$ system in Fig. 2.3, in which the generalized coordinates chosen are $\{q\} = \{X, \theta\}$.

We see that the pendulum's articulation is in movement, in fact, a complex and still indeterminate movement. An actual displacement of mass M of the block is consistent with the constraint it is attached to (i.e., is perpendicular to the normal force $\vec{N}$). On the other hand, the real displacement of the mass m of the pendulum is not perpendicular to the force of constraint $\vec{T}$, the tension due to the string. If we were to use a reasoning identical to that of Sect. 2.1, we would be in trouble, because $\vec{T}$ would not be eliminated. The solution found is to consider the so-called virtual (i.e., not real) displacements, which are (i) infinitesimal; (ii) consistent with the constraints *it is related to*, and; (iii) carried out with fixed t. Second and third conditions allow the virtual displacement to be taken as being perpendicular to the constraint force. However, to reproduce scalar equations of motion in this way (at least in cases where $L = T - V$, see Sect. 2.5.2), we need to write the virtual displacements $\delta \vec{r}_i$ in terms of n' independent generalized coordinates,

$$\delta \vec{r}_\alpha = \sum_{k=1}^{n'} \frac{\partial \vec{r}_\alpha}{\partial q_k} \delta q_k, \tag{2.20}$$

where we made $\delta t = 0$, consistently with the definition of virtual displacements for fixed t. In this equation, δq_k are the corresponding virtual displacements of the generalized coordinates. Why n' instead of n? Because n' represents de number of generalized coordinates related to a particular constraint, not all of them, see rule (ii) above.

To exemplify, let us get back to the system in Fig. 2.3, using capital letters for the block and lower-case letters for the pendulum. As $\vec{R} = X \vec{i}$, the virtual displacement

$$\delta \vec{R} = \frac{\partial \vec{R}}{\partial X} \delta X + \frac{\partial \vec{R}}{\partial \theta} \delta \theta = \delta X \vec{i} \tag{2.21}$$

of mass M, which only depends on X, is perpendicular to the force constraint $\vec{N}$.

The same does not happen with the actual displacement $d\vec{r}$ of mass m. In fact, not only x and y are linked by the constraint equation

$$x^2 + y^2 = l^2, \tag{2.22}$$

but also x depends on X through

$$x = X + l sen\theta. \tag{2.23}$$

On the other hand, being $\vec{r} = (X + lsen\theta)\vec{i} - lcos\theta \vec{j}$, we have,

$$d\vec{r} = \frac{\partial \vec{r}}{\partial X} dX + \frac{\partial \vec{r}}{\partial \theta} d\theta + \frac{\partial \vec{r}}{\partial t} dt = dX \vec{i} + l(cos\theta \vec{i} + sen\theta \vec{j})d\theta, \tag{2.24}$$

because, since $\vec{r}$ does not depend explicitly on t, $\dfrac{\partial \vec{r}}{\partial t}$ vanishes. In turn, the first term is that responsible for $d\vec{r}$ being different of $\delta \vec{r}$, that is, it is responsible for the pendulum hinge be moving horizontally. That is why we must sum over n' instead of n in Eq. (2.20). This equation leads then to,

$$\delta \vec{r} = \frac{\partial \vec{r}}{\partial \theta} \delta \theta = l(cos\theta \vec{i} + sen\theta \vec{j})\delta\theta, \tag{2.25}$$

in which we used rule (ii) to eliminate the term on X of the summation. Now, we easily check that $\delta \vec{r} \cdot \vec{T} = 0$. The next example illustrates this point further.

Example 2.1 In Fig. 2.3, consider the block forced to uniform motion in the x direction with velocity u so that $X = ut$ is no longer a generalized coordinate.

While $d\vec{r} = ut\,\vec{i} + \dfrac{\partial \vec{r}}{\partial \theta}d\theta = ut\,\vec{i} + (l\cos\theta\,\vec{i} + l\sin\theta\,\vec{j})d\theta$ is not orthogonal to $\vec{T}$, $\delta\vec{r}$ will be, since the ut term vanish as demands the (iii) condition.

We will use the concepts of this section to discuss the equivalence of Newtonian and Lagrangian theories in the advent of constraints.

2.3 Hamilton's Principle in Generalized Coordinates

In case the system in study involve constraints, Hamilton's principle does not straightly yields Lagrange's equations because the Cartesian coordinates are no longer independent. But once we work with independent generalized coordinates this obstacle is avoided and things works softly as in Chap. 1. We can thus announce Hamilton's principle in generalized coordinates:

Of all possible trajectories of a system of n degrees of freedom, in configuration space, between fixed points $q(t_1)$ and $q(t_2)$, will be effectively followed the one that minimizes the action S,

$$\delta S = \delta \int_{t_1}^{t_2} L(q, \dot{q}, t)dt = 0. \tag{2.26}$$

The difference to the statement of Hamilton's principle in the previous chapter is that we now have to calculate δS with respect to n independent variations δq_k, as shown in Fig. 2.4. Minimizing the action S,

$$\delta S = \int_{t_t}^{t_2} \delta L(q, \dot{q}; t)dt = \int_{t_t}^{t_2} \sum_{k=1}^{n}[\frac{\partial L}{\partial q_k}\delta q_k + \frac{\partial L}{\partial \dot{q}_k}\delta \dot{q}_k]dt = 0. \tag{2.27}$$

By a procedure totally similar to that of Sect. 1.2, we arrive at

$$\delta S = \int_{t_t}^{t_2} dt \sum_{k=1}^{n}[\frac{\partial L}{\partial q_k} - \frac{d}{dt}\frac{\partial L}{\partial \dot{q}_k}]\delta q_k = 0. \tag{2.28}$$

Due to the presence of constraints, the 3N Cartesian coordinates of the particles in the system will not be independent. In turn, the q_k are independent of each other, so that their arbitrary variations δq_k also are. So the condition $\delta S = 0$ leads us to the well-known Lagrange's equations in generalized coordinates,

$$\frac{d}{dt}\frac{\partial L}{\partial \dot{q}_k} - \frac{\partial L}{\partial q_k} = 0, \quad k = 1, \ldots, n. \tag{2.29}$$

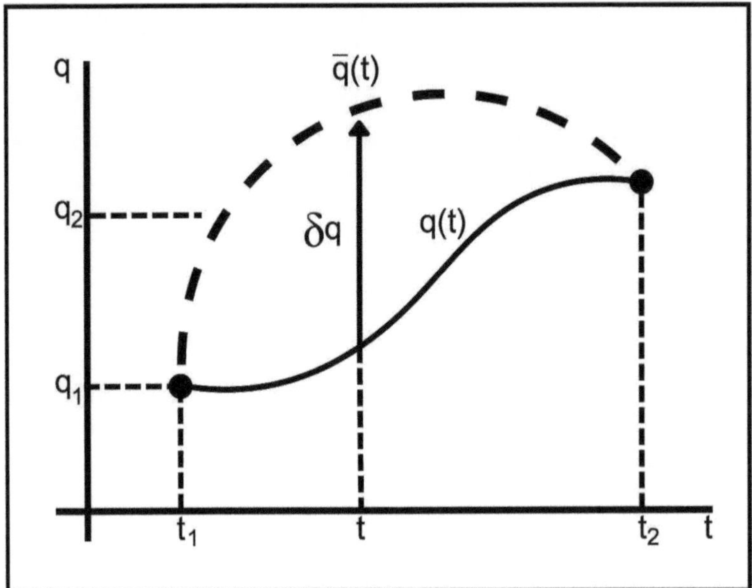

Fig. 2.4 Trajectory variation in configuration space

We normally start writing the Lagrangian using Cartesian coordinates. In order to use Lagrange's equations, we need to identify the holonomic constraints and their number p, so as to choose n independent coordinates, recall Eq. (2.13). In the case of holonomic non-rheonomic constraints, no further reference to them is necessary. The normal procedure for obtaining the Lagrangian in generalized coordinates is to write it first in Cartesian coordinates and use Eqs. (2.16) and (2.17) by direct substitution,

$$L(q, \dot{q}, t) = L[\overrightarrow{r}(q, \dot{q}, t), \dot{\overrightarrow{r}}(q, \dot{q}, t), t]. \tag{2.30}$$

2.3.1   *Invariance of Lagrange's Equations Under Point Transformations*

A change of generalized coordinates of the type

$$Q_j = Q_j(q_k), \quad j, k = 1, n \tag{2.31}$$

is called a *point transformation* in the configuration space. From the considerations made so far, it seems obvious that *Lagrange's equations are invariant under a point transformation.*

This means that, once Lagrange's equations (2.29) are valid for the q_k, the equations

$$\frac{d}{dt}\frac{\partial L}{\partial \dot{Q}_j} - \frac{\partial L}{\partial Q_k} = 0, \quad j = 1, \ldots, n. \tag{2.32}$$

are also fulfilled.

An informal argument is what follows: Let q_1, q_2 be two generalized coordinates of a mechanical system ($n = 2$) and $Q = f(q)$ a point transformation. For each q there is one and only one Q, that is, the set $\{Q\}$ is eligible as generalized coordinates, provided they are independent. Now assume that $Q_2 = g(Q_1)$. In consequence, $f(q_2) = g[f(q_1)] = h(q_1)$. This would mean a constraint relation between q_1 and q_2, which can not exist since, by hypothesis, q_1 and q_2 are independent. Hence, the analogous relation g between Q_1 and Q_2 can not exist either; they are also independent.

Exercise Prove the invariance directly by substitution of the point transformation to Lagrange's equation for q_k.

2.4 Newton's Second Law from Lagrange's Equations

Now, consider the case where the generalized coordinates are Cartesian, $x_{\alpha j}$, in the absence of constraints. Being

$$L = \sum_{\alpha=1}^{N}\sum_{j=1}^{3} \frac{m_\alpha}{2}\dot{x}_{\alpha j}^2 - V(x_{\alpha j}), \tag{2.33}$$

Lagrange's equations produce, for a specific pair (α, j),

$$\frac{d}{dt}\frac{\partial L}{\partial \dot{x}_{\alpha j}} = m_\alpha \ddot{x}_{\alpha j}, \quad \frac{\partial L}{\partial x_{\alpha j}} = -\frac{\partial V}{\partial x_{\alpha j}} = F_{\alpha j} \;\Rightarrow\; m_\alpha \ddot{x}_{\alpha j} = F_{\alpha j}, \tag{2.34}$$

that is, they recover the components of the Newtonian law of motion.

What happens if we use non-cartesian vector coordinates? We will illustrate this point with an example, taking the Lagrangian of a particle in plane motion with polar coordinates (r, φ), under the action of a central force $\vec{F} = -\vec{\nabla} V(r)$ (later, we will show how this plane movement is possible despite no constraint is involved),

$$L = \frac{m}{2}(\dot{r}^2 + r^2\dot{\varphi}^2) - V(r). \tag{2.35}$$

Exercise Obtain this Lagrangian starting with cartesian coordinates.

Lagrange's equations for the generalized coordinates r and φ are, respectively,

$$m\ddot{r} - mr\dot{\varphi}^2 + \frac{dV}{dr} = 0 \quad e \quad mr\ddot{\varphi} + 2mr\dot{r}\dot{\varphi} = 0. \tag{2.36}$$

These are the same equations of motion obtained when we write $\overrightarrow{F} = m\overrightarrow{a}$ in polar coordinates, which the reader can easily verify.

As a general conclusion, for a system of unconstrained particles Newton's and Lagrange's equations are equivalent. In consequence, give up the idea, if it has ever crossed your mind, that Hamilton's principle shows particles "choosing" their paths. Approaching a problem in classical mechanics either by the integral Hamilton's principle or by the Newtonian method of building the particle's trajectory step by step, are equivalent ways of describe the movement (never despising, on the other hand, that the freedom in choosing $L \neq T - V$ allows possibilities not allowed by using Newton's mechanics).

However, in the presence of constraints these features changes. For example, if we set $r = R$ fixed, the Lagrangian becomes simply $L = \frac{m}{2}R^2\dot{\varphi}^2$ and Lagrange's equation for φ yields

$$\frac{d}{dt}(mR^2\dot{\varphi}) = 0, \tag{2.37}$$

which expresses the conservation of angular momentum of the particle and does not, of course, have the form $\overrightarrow{F} = m\overrightarrow{a}$, because now $\overrightarrow{a} = R\ddot{\varphi}\hat{\varphi}$.

It should finally be recalled that, when using Cartesian coordinates in a problem with constraints, we accept a redundancy in the number of coordinates needed to describe the motion. In other words, constraint forces are explicit components of the resultant force on each particle, F_{ik}.

2.5 Lagrange's Equations from Newton's Second Law

As we have seen that Lagrange's equations reproduce Newton's second law, a pertinent question would be whether the reciprocal is true, that is, can we start from the second law and get to Lagrange's equations? Initially, it is necessary to understand that this approach would imply a restriction of the applicability of Lagrangian mechanics to cases where Newtonian mechanics applies. Particularly, the examples involving the theory of relativity would be excluded, but not only them. In fact, we will see that all cases in which L differs from $T - V$, even those embodied in Newtonian mechanics, will also be excluded. Symptomatically, such approach, known as *D'Alembert's Principle*, makes use the form $L = T - V$ for the Lagrangian of conservative systems, despite written in non-vector generalized coordinates.

2.5.1 *Absence of Constraints*

Instead of starting with vector Newton's second law, which is more appropriate for problems involving constraints (and will be done in the next section), we have chosen here a theoretically proper scalar approach. We define the *generalized moment* (also called canonical moment) associated with coordinate q_k by

$$p_k = \frac{\partial L}{\partial \dot{q}_k}. \tag{2.38}$$

In this chapter, the potential energy V does not depend on velocities, so the above definition leads to

$$p_k = \frac{\partial T}{\partial \dot{q}_k} = \frac{\partial}{\partial \dot{q}_k}\left(\frac{m}{2}\sum_{j=1}^{3}\dot{x}_j^2\right) = \sum_{j=1}^{3} m\dot{x}_j \frac{\partial \dot{x}_j}{\partial \dot{q}_k}. \tag{2.39}$$

Without loss of generality, only one particle is considered here. Our approach to arrive to Lagrange's equations using Newton's law is to search for the time derivative of p_k. We consider the components of the transformation equations (2.16),

$$x_j = x_j(q_k, t), \tag{2.40}$$

such that, in terms of the generalized velocity components (2.17),

$$\dot{x}_j = \sum_{k=1}^{n} \frac{\partial x_j}{\partial q_k}\dot{q}_k + \frac{\partial x_j}{\partial t}, \tag{2.41}$$

from which we can extract,

$$\frac{\partial \dot{x}_j}{\partial \dot{q}_k} = \frac{\partial x_j}{\partial q_k}. \tag{2.42}$$

Taking these results into Eq. (2.39), we get

$$p_k = \sum_{j=1}^{3} m\dot{x}_j \frac{\partial x_j}{\partial q_k} \quad\Rightarrow\quad \dot{p}_k = \sum_{j=1}^{3}\left(m\ddot{x}_j \frac{\partial x_j}{\partial q_k} + m\dot{x}_j \frac{d}{dt}\frac{\partial x_j}{\partial q_k}\right), \tag{2.43}$$

or

$$\dot{p}_k = \sum_{j=1}^{3} m\ddot{x}_j \frac{\partial x_j}{\partial q_k} + \sum_{j=1}^{3}\sum_{l=1}^{n} m\dot{x}_j \frac{\partial^2 x_j}{\partial q_l \partial q_k}\dot{q}_l + \sum_{j=1}^{3} m\dot{x}_j \frac{\partial^2 x_j}{\partial t \partial q_k}. \tag{2.44}$$

The first term on the right is the generalized force Q_k,

$$Q_k = \sum_{j=1}^{3} F_j \frac{\partial x_j}{\partial q_k} = -\sum_{j=1}^{3} \frac{\partial V}{\partial x_j} \frac{\partial x_j}{\partial q_k} = -\frac{\partial V}{\partial q_k}, \qquad (2.45)$$

where F_j, using Newton's law at this point, is the jth component of force on the particle. The next two terms correspond to $\dfrac{\partial T}{\partial q_k}$,

$$\frac{\partial T}{\partial q_k} = \frac{\partial}{\partial q_k}\left(\frac{m}{2} \sum_{j=1}^{3} \dot{x}_j^2\right) = \sum_{j=1}^{3} m\dot{x}_j \frac{\partial \dot{x}_j}{\partial q_k} = \sum_{j=1}^{3} m\dot{x}_j \frac{\partial}{\partial q_k}\left(\sum_{l=1}^{n=3} \frac{\partial x_j}{\partial q_l}\dot{q}_l + \frac{\partial x_j}{\partial t}\right),$$
$$(2.46)$$

as concluded by comparison with Eq. (2.44). That is, when we move from Cartesian to generalized coordinates, the kinetic energy T can become dependent not only on generalized velocities, but also on generalized coordinates (see, for example, Eq. (2.35)). Finally we get

$$(\dot{p}_k =)\frac{d}{dt}\frac{\partial T}{\partial \dot{q}_k} = -\frac{\partial V}{\partial q_k} + \frac{\partial T}{\partial q_k}. \qquad (2.47)$$

This equation looks like suggesting that we consider a function $L = T - V$ as a generator of the movement, since as V does not depend on $\dot{q}_k$ in conservative systems, it results that,

$$\frac{d}{dt}\frac{\partial(T - V)}{\partial \dot{q}_k} - \frac{\partial(T - V)}{\partial q_k} = 0, \quad k = 1, 2, 3. \qquad (2.48)$$

We then arrived to Lagrange's equation.

Some considerations are pertinent. The dependency of T on q_k arises when changing from Cartesian to generalized coordinates, regardless of the existence of constraints. This conclusion can be extended to unconstrained many-particle systems, with $n = 3N$. If the (or some) generalized coordinates are the Cartesian coordinates themselves, then the dependence on q_k disappears. The "suggestion" $L = T - V$ came out independently of the existence of constraints. If there are constraints to the motion of the system, a different approach is needed.

2.5.2   Presence of Constraints: D'Alembert's Principle

If the system under consideration is subject to constraints, the unknown constraint forces $\overrightarrow{f}_\alpha$ enter Newtonian dynamics. To make the corresponding formulation rid of them was probably the first motivation for the Lagrangian theory. For that, it was necessary to resort to a dynamical version of the principle of virtual works. This principle, very well established in statics, rely on multiplying each constraint force by the correspondent virtual dislocation (the *virtual work*, which results equal to zero as justified in Sect. 2.2.1).

But in order to apply it to dynamics it is necessary to introduce the mathematical artifice of virtual work of the "force" $\dot{\vec{P}}_\alpha$, that is, $\dot{\vec{P}}_\alpha \cdot \delta \vec{r}_\alpha$.[3] D'Alembert's principle of virtual works, presents itself, then, in the form

$$\sum_{\alpha=1}^{N}(\dot{\vec{P}}_\alpha - \vec{F}_\alpha) \cdot \delta \vec{r}_\alpha = 0, \quad \vec{f}_\alpha \cdot \delta \vec{r}_\alpha = 0, \tag{2.49}$$

where $\vec{F}_\alpha$ indicates only applied forces. This principle then allows the elimination of constraint forces $\vec{f}_\alpha$ by scalar multiplying them by $\delta \vec{r}_\alpha$, the virtual displacements defined by Eq. (2.20). As seen previously (Sect. 2.2.1), these virtual displacements are chosen at fixed t, so as to be orthogonal to the constraint forces $\vec{f}_\alpha$.

Once we take $3 > n \geq 1$ (one or two constraint equations), we do not loose generality by considering only one particle, $\alpha = 1$. For the deduction, we will need the result,

$$\frac{d}{dt}\frac{\partial x_j}{\partial q_k} = \sum_{l=1}^{n} \frac{\partial}{\partial q_l}(\frac{\partial x_j}{\partial q_k})\dot{q}_l + \frac{\partial}{\partial t}(\frac{\partial x_j}{\partial q_k}) = \frac{\partial}{\partial q_k}(\sum_{l=1}^{n} \frac{\partial x_j}{\partial q_l}\dot{q}_l + \frac{\partial x_j}{\partial t}) = \frac{\partial \dot{x}_j}{\partial q_k}, \tag{2.50}$$

where we used Eq. (2.41) in the last passage. Taking expression (2.20) of the virtual displacements (here $n' - n$) into D'Alembert's equation (2.49), with $\alpha = 1$,

$$\sum_{k=1}^{n}(\dot{\vec{P}} - \vec{F}) \cdot \frac{\partial \vec{r}}{\partial q_k}\delta q_k = \sum_{k=1}^{n}\sum_{i=1}^{3}(\dot{P}_i\frac{\partial x_i}{\partial q_k} - F_i\frac{\partial x_i}{\partial q_k})]\delta q_k = 0. \tag{2.51}$$

The first term in parentheses on the right, considering Eq. (2.43) with $P_i = m\dot{x}_i$, leads to

$$\sum_{i=1}^{3}\dot{P}_i\frac{\partial x_i}{\partial q_k} = \dot{p}_k - m\sum_{i=1}^{3}\dot{x}_i\frac{d}{dt}\frac{\partial x_i}{\partial q_k}) = \dot{p}_k - m\sum_{i=1}^{3}\dot{x}_i\frac{\partial \dot{x}_i}{\partial q_k} = \dot{p}_k - \frac{\partial T}{\partial q_k}, \tag{2.52}$$

where we used Eq. (2.50). Now the second term corresponds to the generalized force (2.45), so we get

$$\sum_{k=1}^{n}(\dot{p}_k - \frac{\partial T}{\partial q_k} + \frac{\partial V}{\partial q_k})\delta q_k = 0. \tag{2.53}$$

---

[3] Here, the linear momentum is represented with a capital letter so there be no confusion between its components and the generalized moments (2.38).

If there are equations of holonomic constraints involving the Cartesian coordinates, the δx_j components of the virtual displacements are not independent. Here, however, the δq_k are arbitrary and independent of each other, which allows zeroing their coefficients in the equation above. Remembering that $\dot{p}_k = \dfrac{d}{dt}\dfrac{\partial T}{\partial \dot{q}_k}$ and that V does not depend on the generalized velocities, we obtain Lagrange's equations identical to (2.48), except that they will here be limited in number to $n < 3$,

$$\frac{d}{dt}\frac{\partial(T-V)}{\partial \dot{q}_k} - \frac{\partial(T-V)}{\partial q_k} = 0, \quad k = 1, n. \tag{2.54}$$

Decreasing the number of degrees of freedom of the particle is a consequence of the elimination of constraint forces. The generalization for a system of N particles, p constraint equations and $n = 3N - p$ degrees of freedom is immediate, according to Eq. (2.54).

2.5.3   The Lagrangian for Planar Motion of Rigid Bodies

The kinetic energy of rigid-bodies in planar rotation (that is, the axis of rotation keeps its direction fixed), is given by $T_{rot} = \frac{1}{2}I\omega^2 = \frac{1}{2}I\dot{\varphi}^2$, where I is its moment of inertia relative to the axis of rotation and φ the angle of rotation. It must be added to the translational kinetic energy of particles and bodies in the Lagrangian, of course. The angular generalized coordinate φ is enough for this kind of rotating body, but there might be the increment of one linear degree of freedom, in case of rolling with slipping.

Example 2.2 A block of mass m is attached to the axis of a disk of same mass m and moment of inertia $I = \frac{1}{2}mR^2$, as shown in Fig. 2.5.

Assuming that the spring is unstretched in the situation where the CM of the bodies coincide and that there is no slipping of the disk, we can easily approach the dynamic problem using Lagrange equations.

For translation we consider the coordinates x_1 and x_2. In turn, $z_1 = z_2 = 0$, $y_1 = y_2 = R$ are 4 equations of constraint, but here, again, formula (2.13) fails, because one more degree of freedom and a further constraint equation must be added to account for the disk rotation. So, the right math is $n = 7 - 5 = 2$. The angle φ describes the planar rotation of the disk and pure rolling condition, $\dot{x}_2 = R\dot{\varphi}$, when integrated with $\varphi_0 = 0$, leads to the expected extra constraint equation $x_2 - R\varphi = 0$, so we really have two degrees of freedom. We then choose x_1 and θ as generalized coordinates. The Lagrangian is

$$L = \frac{m}{2}\dot{x}_1{}^2 + \frac{m}{2}\dot{x}_2{}^2 + \frac{I}{2}\dot{\varphi}^2 - \frac{k}{2}(x_2 - x_1)^2 = \frac{m}{2}\dot{x}_1{}^2 + \frac{3m}{4}R^2\dot{\varphi}^2 - \frac{k}{2}(R\varphi - x_1)^2. \tag{2.55}$$

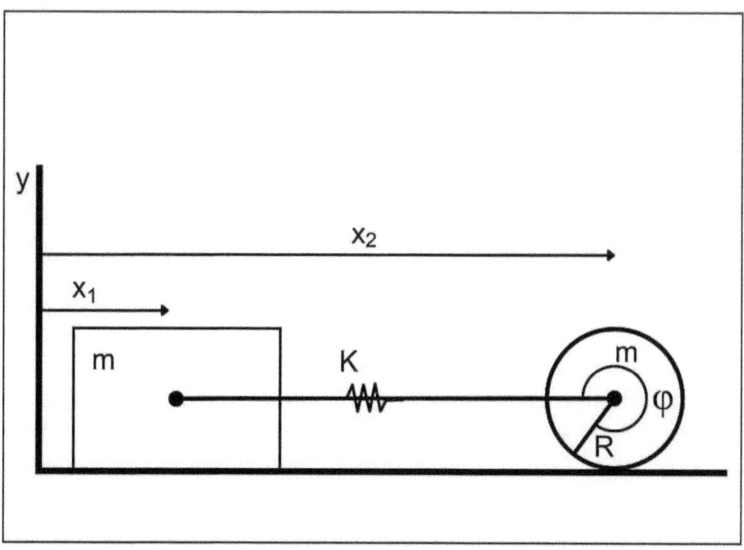

Fig. 2.5 System of a block and a disk linked by a string

The equations of motion can be easily obtained,

$$\ddot{x}_1 + \frac{k}{m}(x_1 - R\varphi) = 0 \quad e \quad \ddot{\varphi} + \frac{2k}{3mR}(R\varphi - x_1) - 0. \tag{2.56}$$

They produce two decoupling situations: Taking $\varphi = 0$ (fixed disk) in the first yields the equation of a harmonic oscillator for body 1. In turn, $x_1 = 0$ (block fixed), produces

$$\ddot{\varphi} + \frac{2k}{3m}\varphi = 0, \tag{2.57}$$

meaning that the disk will go forth and back also in a harmonic way.

2.6 Moving Constraints and Reference Frames

In the solution of problems involving accelerated constraints, the so-called fictitious forces appear in Lagrange's equations. The beginner can get confused with this because what he normally does could be seen as a "simple" choice of generalized coordinates, replacing the original Cartesian ones, accompanying the movement of the constraint. The fact is that Hamilton's principle does not impose the specification of an inertial frame of reference to study the dynamics of mechanical systems. This information will be contained in the Lagrangian, through its kinetic energy term. In turn, it is normal to think that we refer to an inertial system when we state this principle. In fact, this section could be located at Chap. 1, in view of its fundamental character, but its connection with accelerated constraints advises treating it here.

To approach this important point, let us check how the Lagrangian changes in the case of a rotating system with angular velocity $\vec{\omega}$ constant (which corresponds Earth's rotation, for example). There is no loss in considering a single free particle, since the relevant effects affect just its kinetic energy, or even in using Cartesian coordinates, since an eventual switch to curvilinear coordinates can be made after changing the axis system.

The Cartesian coordinates of a particle in the inertial frame does not depend explicitly on t, neither does T or, in fact, L (since V depends only on relative distances). T may become explicitly time-dependent in consequence of the choice of a minimal set of generalized coordinates relative to the moving frame. As will be shown later here, see Eqs. (2.105–2.108), or is shown in basic textbooks, this explicit dependence $T(t)$ is connected to the appearance of extra (non-quadratic in the generalized velocities) terms in it. These extra terms must be responsible for the appearance of non-inertial effects in the common application of Lagrange's equations to problems involving accelerated constraints, but with no explanation of from where they really come.

Let us make formal the discussion in order to get a deeper understanding of the topic. We take the axes such that the non-inertial system, with coordinates (x, y, z), rotate around the common z axis (for simplicity), of the inertial system, of coordinates (X, Y, Z), see Fig. 2.6. We thus approach the problem of moving from an inertial frame to a rotating frame with constant angular velocity ω along $z = Z$,

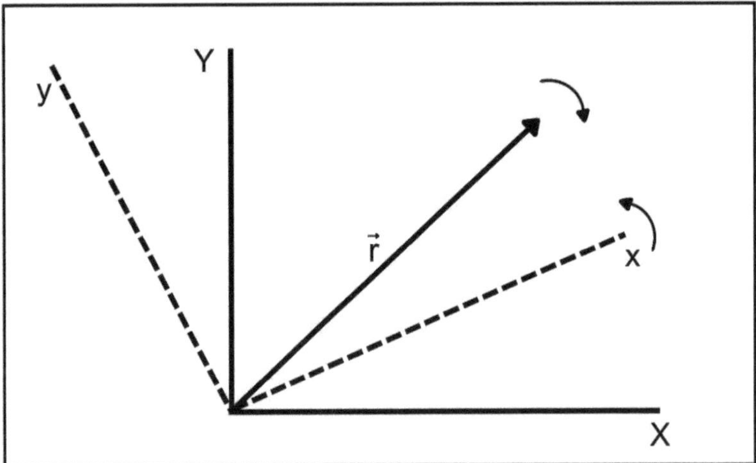

Fig. 2.6 Fixed (X, Y) and rotating (x, y) axes. Rotating axes is equivalent to counter-rotate the vector

while describing the motion of a single particle. Again, the potential energy is not affected by this transformation and need not be taken into account. The motion of a free particle contains all we need for the discussion. Also, the frames will share the same origin (the extension to translational relative motion of the frames is made in the next Example).

The Lagrangian in the inertial frame coordinates is simply $L = \dfrac{m}{2}(\dot{X}^2 + \dot{Y}^2 + \dot{Z}^2)$. It will be written in terms of the x, y, z coordinates in the rotating frame, which, for now, seems like only a change of coordinates. For counter-clockwise rotation, the two sets of coordinates are related by

$$X = x\cos\omega t - y\,sen\,\omega t \tag{2.58}$$

$$Y = x\,sen\,\omega t + y\cos\omega t$$

$$Z = z$$

Then, the Lagrangian written in terms of coordinates (x, y, z) is obtained as,

$$L = \frac{m}{2}(\dot{X}^2 + \dot{Y}^2 + \dot{Z}^2) = \frac{m}{2}[(\dot{x}^2 + \dot{y}^2 + \dot{z}^2) + \omega^2(x^2 + y^2) + \omega(x\dot{y} - y\dot{x})]. \tag{2.59}$$

This Lagrangian is still relative to the inertial frame, despite being now a function of generalized coordinates and velocities relative to the accelerated frame. It will produce the proper equations of motion as known. On the other hand, the temptation to identify a Lagrangian L^{acc} relative to the accelerated frame arises by inspection of the right hand side of this equation. As we identify $T^{acc} = \dfrac{m}{2}[(\dot{x}^2 + \dot{y}^2 + \dot{z}^2)]$ as the kinetic energy at the accelerated frame, the remaining terms can be interpreted as fictitious potentials in this frame. Everything would happen as if an observer in the accelerated frame added, ad hoc, fictitious potential energies to T^{acc} to achieve a proper solution for the problem. This proposition looks equivalent to the way this subject is treated in Newtonian mechanics, e.g., by the introduction of fictitious forces in order to second Newton's law to work in accelerated reference frames. Moreover, L^{acc} is obtained in terms of a valid set of generalized coordinates and velocities (and time, eventually), so there is no doubt that application of Hamilton's principle would lead to Lagrange's equations for the new coordinates *in the accelerated frame.*

Let us check the meaning of the "extra" terms in L. The second term in Eq. (2.59) is easily identified as a fictitious centrifugal potential energy, since it yields the centrifugal force through,

$$V_{cf} = -\frac{1}{2}m\omega^2(x^2 + y^2), \quad \vec{F}_{cf} = -\vec{\nabla}V_{cf} = m\omega^2\vec{r}. \tag{2.60}$$

The last term must correspond to the fictitious Coriolis potential energy,

$$U_{cor} = m\omega(x\dot{y} - y\dot{x}). \tag{2.61}$$

In fact, being a velocity-dependent potential, the analogy with the electromagnetic case yields,

$$\vec{F}_{cor} = -\vec{\nabla} U_{cor} + \frac{d}{dt}\frac{\partial U_{cor}}{\partial \vec{v}} = -2m\omega(\dot{y}\,\vec{i} + \dot{x}\,\vec{j}), \tag{2.62}$$

which is really identified as the fictitious Coriolis force of Newtonian mechanics.

In other words, moving coordinates from the inertial to the non-inertial frames, the last represented by x, y here, automatically introduces the corresponding terms in L and in the equations of motion, without explicit reference to fictitious forces. On the other hand, Eq. (2.59) shows that a non-inertial observer would have to include the fictitious potential energies in order to obtain the correct equations of motion in the non-inertial frame. The standard procedure, however, does not demand the inclusion of ad hoc terms in the Lagrangian *beyond* the choice of generalized coordinates associated to the accelerated frame. This choice does it automatically, as shown in examples that follow.

In passing, the consideration of a time dependence of ω would bring an extra fictitious potential energy term, normally dismissed, but no further difficulties.

Concerning now accelerated translation of a reference frame, an analogous procedure leads to the general expression of the Lagrangian of a single particle in 1D motion. In an inertial frame the particle have coordinate X, being x its coordinate in an accelerated frame, with constant acceleration a, and $x' = \frac{1}{2}at^2$ is the distance of the origins of the two frames. Being $X = x' + x$, time derivative yields $\dot{X} = \dot{x} + at$ so that the Lagrangian becomes,

$$L = \frac{1}{2}m\dot{X}^2 = \frac{1}{2}m\dot{x}^2 + \frac{1}{2}ma^2t^2 - max, \tag{2.63}$$

where we considered the origins of the two frames coincident at $t = 0$, so that $x = -\dot{x}t$.

Again, the Lagrangian has been written in the inertial frame but using the generalized coordinate x linked to the accelerated frame.

Thus, since the first term is the kinetic energy in the accelerated frame, there is a fictitious *recoil* potential energy $V_{rec} = max$ in this frame. The recoil force is then

$$V_{rec} = max, \quad \vec{F}_{rec} = -\nabla V_{rec} = -m\vec{a}, \tag{2.64}$$

as it should be.

As in the previous case, this development will automatically be taken into account when we choose a generalized coordinate moving along with the accelerated constraint, so that no explicit reference needs to be made to any reference frame or to fictitious potentials. In general, there are no advantages in using the Lagrange formulation to handle problems with moving frames, but when a problem involves accelerated constraints there is no way to avoid these considerations. In an inertial frame of reference, in the case of accelerated constraint, the kinetic energy T includes terms introduced by the equations of rheonomic constraints, through the transformation equations, which express, in the equations of motion, the action of constraint forces on the system. A referential linked to the constraint is necessarily non-inertial and the Lagrangian in this frame must contain fictitious potential energy terms corresponding to the fictitious forces, which replace the kinetic terms of the inertial frame. No matter which reference frame we are concerned to, the equations of motion will necessarily be the same whenever the same generalized coordinates are used. Examples are useful for establishing these concepts.

Exercise A body in the bottom of a car is accelerated horizontally, with acceleration a, by static friction. Using $X = x + \frac{1}{2}at^2$, show that the resulting Lagrangian in terms of x and $\dot{x}$ yields the proper equation of motion. Show that the Lagrangian L^{acc}, written by the non-inertial observer, is equivalent to the previous, differing only by $\frac{1}{2}a^2t^2$. Check that the resulting equation of motion is the same.

Example 2.3 Consider the case of a simple pendulum inside a car that has acceleration a in the direction x, see Fig. 2.7.

We start handling this problem from the inertial reference frame but using the angle θ as our only generalized coordinate ($n = 1$) linked to the constraint. The rheonomic constraint equation that justifies this choice is $X - l sen\theta - \frac{1}{2}at^2 = 0$. According to the figure, but disregarding the constant height of the pendulum pivot, the transformation equations are

$$X = \frac{1}{2}at^2 + l sen\theta, \quad y = -l cos\theta \, . \tag{2.65}$$

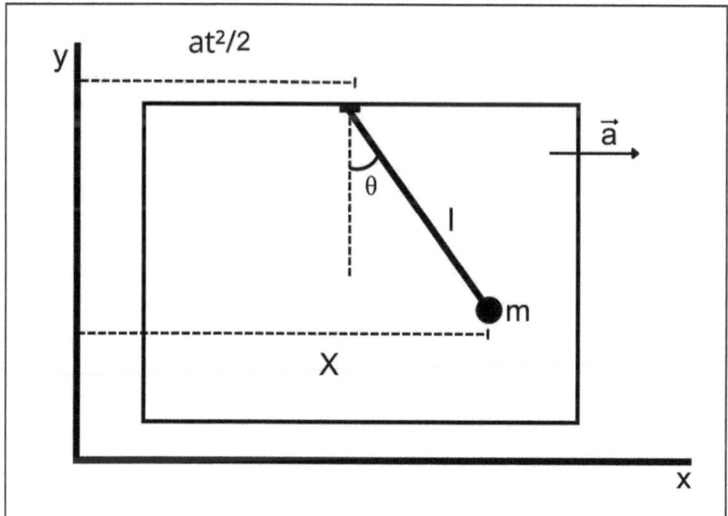

Fig. 2.7 Pendulum inside a moving car

The Lagrangian is easily obtained as

$$L = \frac{m}{2}(a^2t^2 + l^2\dot{\theta}^2 + 2alt\dot{\theta}cos\theta) + mglcos\theta. \tag{2.66}$$

The equation of motion is then

$$\ddot{\theta} + \frac{g}{l}sen\theta + \frac{a}{l}cos\theta = 0, \tag{2.67}$$

which differs from the simple pendulum case by the last term on the left. This term portrays the effect of the constraint on the pendulum. We conclude that, contrary to fixed constraints, *an accelerated constraint does act dynamically on the system!*

Moving now to a non-inertial frame linked to the car, the transformation equation for X will no longer contain the term $\frac{1}{2}at^2$ and the kinetic energy will simply be $T = \frac{m}{2}l^2\dot{\theta}^2$, but now we have to add to the potential energy V the term corresponding to the fictitious force $-m\overrightarrow{a}$, that is, $V^{acc} = max = malsen\theta$. The Lagrangian becomes, then,

$$L^{acc} = \frac{m}{2}l^2\dot{\theta}^2 + mglcos\theta - malsen\theta. \tag{2.68}$$

As already advanced, this Lagrangian produces the same equation of motion (2.67), where the last term results from the absolute character of the accelerated motion of the car. Notice that the Lagrangians (2.66) and (2.68) are equivalent, as they differ by

$$L - L^{acc} = \frac{d}{dt}(\frac{1}{6}a^2t^3 + maltsen\theta). \tag{2.69}$$

The equal solutions is not, then, a coincidence and should not be seen with strangeness. After all, in both reference frames the same generalized coordinate θ, whose temporal evolution is independent of of the frame used.

Example 2.4 Another enlightening problem is that of a bead that runs free of friction along a rod, which rotates with constant angular velocity ω, on a horizontal plane, as shown in Fig. 2.8.

The relevant vector coordinates are r and θ, but given the constraint equation

$$\theta - \omega t = 0 \tag{2.70}$$

we use only r as an independent generalized coordinate. In the inertial frame there is no potential energy, so that the transformation equations,

$$x = rcos\omega t, \quad y = rsen\omega t \tag{2.71}$$

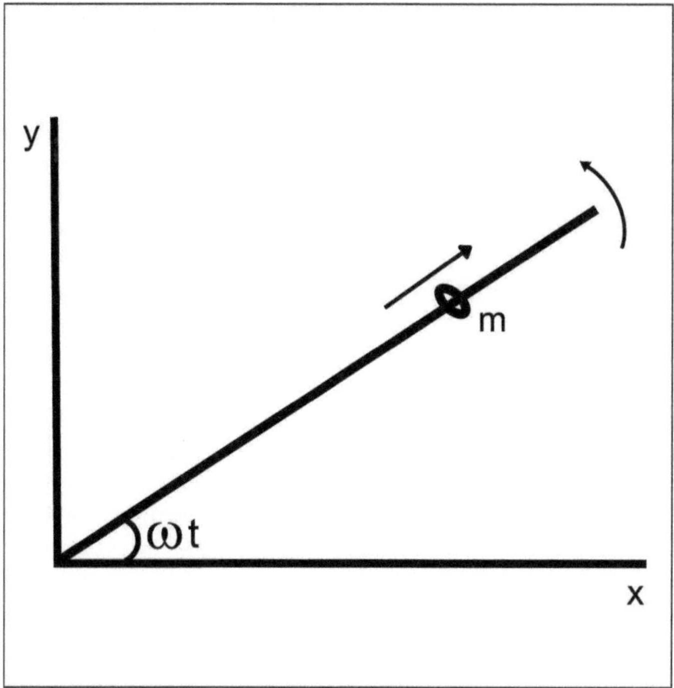

Fig. 2.8 Bead moving along a rotating bar

yield

$$L = T = \frac{m}{2}(\dot{r}^2 + r^2\omega^2),$$ (2.72)

which leads us to the equation of motion

$$m\ddot{r} - m\omega^2 r = 0.$$ (2.73)

In this equation, the second term results from the force that the constraint exerts on the bead, perpendicular to the rod, which forces the particle to "go off the tangent", so to increase r. In fact, let us take a fixed point on the rod that instantly coincides with the position of the bead. This point will execute a circular trajectory, while the bead is forced, by the constraint force, to the tangent direction, that is, it is accelerated radially outward by the constraint force. So, Eq. (2.73) above does not imply moving to a non-inertial reference frame, so there is no sense in calling its second term the "centrifugal force". In fact, if we now move to the non-inertial frame attached to the rod, the kinetic energy will be only $T = \frac{m}{2}\dot{r}^2$, but there will be a centrifugal potential energy $V^{cf} = -\frac{m}{2}r^2\omega^2$, producing the same Lagrangian (2.72) and the same equation of motion (2.73), but now it makes sense calling that term the centrifugal force.

Obtaining the same result in the two, inertial and non-inertial reference frames, can lead to the false impression that the question is irrelevant. However, it is not, as we get some new information. Not only the perpendicular constraint force, but also the (equally perpendicular) Coriolis force are ruled out as the redundant coordinate $\phi = \omega t$ is ignored for this one-degree-of-freedom problem, even if one is treating the problem in the accelerated frame. It is like they "cancel each other out", just the centrifugal force remaining, an obligatory assumption in the accelerated frame. Nevertheless, choosing r as the only generalized coordinate in the inertial frame, solves the problem without these considerations, which stands as another impressive feature of Hamilton's principle.

Constraints explicitly independent of t represent limitations to the configurational space, but once satisfied with the choice of n generalized coordinates, they play no further action on the systems, as we know. On the other hand, the dynamic action of accelerated constraints, in inertial frames, shows up as terms derived from their kinetic energy. In fact, they are actions of the constraint forces, which are taken into account by means of the rheonomic constraint equations. The idea, then, that the constraint forces are eliminated from the mechanical problem by the use of a minimum set of coordinates (n) does not apply to systems subject to accelerated constraints, from the point of view of an inertial reference frame. It is only valid in the non-inertial frame attached to the constraint. Evidently, a constraint can have a more complex movement, with variable acceleration, for example, but these conclusions should be kept.

Example 2.5 Let us now return to the consideration of the system of Example 2.3, but with the condition of the speed $\vec{u}$ of the car ($\vec{u}(0) = 0$) being constant. Note that it is not enough to make $\vec{a} = 0$ in the Lagrangian (2.66) because there, for simplicity, the initial velocity was taken as null. The necessary modification in the derivation comes from the fact that the constraint equation is now

$$X - l sen\theta - ut = 0, \tag{2.74}$$

which corresponds to the first transformation equation, while the second is the same of Eq. (2.65). The new Lagrangian becomes,

$$L = \frac{m}{2}u^2 + \frac{m}{2}l^2\dot{\theta}^2 + mul\dot{\theta}cos\theta + mglcos\theta. \tag{2.75}$$

As happened in the case of the Lagrangians (2.66) and (2.68), the third term here can be written as

$$mul\dot{\theta}cos\theta = \frac{d}{dt}(mulsen\theta). \tag{2.76}$$

The Lagrangian above is then equivalent to that of the simple pendulum with immovable articulation, $u = 0$ (the term $\frac{m}{2}u^2$, constant, has no effect on the equations of motion). This can be verified by deriving the equation of motion,

$$\frac{\partial L}{\partial \theta} - \frac{d}{dt}\frac{\partial L}{\partial \dot{\theta}} = -ml^2\ddot{\theta} + mlu\dot{\theta}sen\theta - mglsen\theta - mlu\dot{\theta}sen\theta \qquad (2.77)$$

or,

$$\ddot{\theta} + \frac{g}{l}sen\theta = 0, \qquad (2.78)$$

that is, the pendulum equation with fixed origin was obtained. It couldn't be different! The reference frames, in this case, are both inertial and the Galilean principle of relativity necessarily manifests itself, which it does through equivalent Lagrangians. Once again, we observe that a fundamental symmetry of nature is contained in the Lagrangian. Other aspects of Galilean invariance will be explored later, after the introduction of conservation theorems.

Exercise Three apparently analogous problems are proposed, which serve to fix the concepts studied in this section, see Figs. 2.3 and 2.9. In all, a simple pendulum is attached to a block that can move horizontally. In the scheme on the left in Fig. 2.9, the block is forced by an external agent to oscillate, so that, in this case, the pendulum is subject to a mobile constraint, whose action on the movement of the pendulum will be taken into account through of the transformation equations. In the others, in Fig. 2.3 and on the right in Fig. 2.9, the motion of the block is not

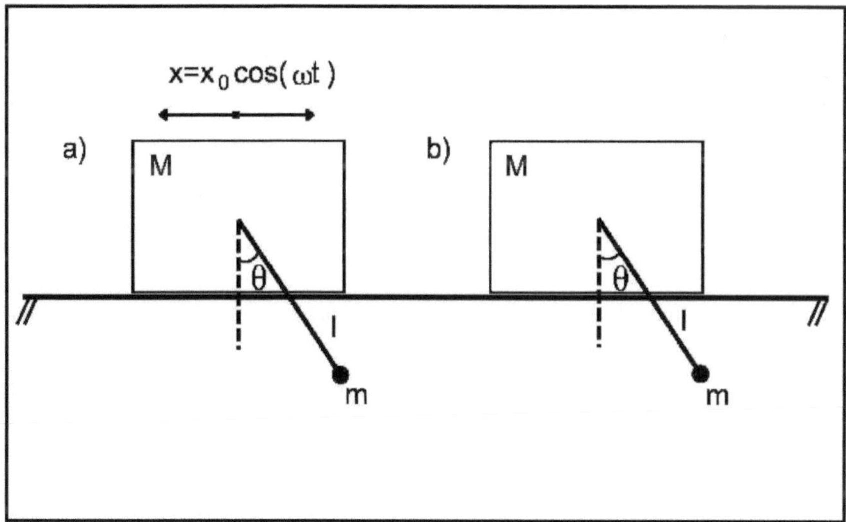

Fig. 2.9 Blocks with coupled pendula

previously known; we no longer have a pendulum subject to a moving constraint but rather two-body, two-degree-of-freedom system, whose motion is coupled. Obtain Lagrange's equations for the three cases and analyse them in terms of the topics discussed in the last section.

2.7 Symmetries of the Lagrangian and Conservation Theorems

2.7.1 Constants of Motion

A quantity $F(q, \dot{q}, t)$ is a *constant of motion* if its total time derivative is zero,

$$\dot{F}(q, \dot{q}, t) = 0 \tag{2.79}$$

It is important to understand the conceptual difference between an universal constant and a constant of motion. For example, the velocity of light c is an universal constant of great importance, but it is not a constant of motion (in the present context, at least), as its value is fixed regardless of the values assumed by the dynamic variables. A constant of motion can assume different constant values, depending on the initial conditions, but keep their values then.

Direct verification of the time invariance of a quantity can be done using the equations of motion. Let us take, for example, the rate of change of the mechanical energy E of a particle in 1D motion under a potential V. We have,

$$E = \frac{m}{2}\dot{x}^2 + V \;\;\Rightarrow\;\; \dot{E} = \frac{m}{2}\dot{x}^2 + V = m\dot{x}\ddot{x} + \frac{dV}{dx}\dot{x} = (m\ddot{x} + \frac{dV}{dx})\dot{x} = 0, \tag{2.80}$$

since $m\ddot{x} = -\dfrac{dV}{dx}$. Likewise, it can be seen that in the presence of a dissipative force, say $-c\dot{x}$, we will have,

$$\dot{E} = -c\dot{x}^2 \neq 0. \tag{2.81}$$

2.7.2 Symmetries and Cyclical Coordinates

There is a remarkable property of the Lagrangian that ties in with fundamental conservation principles of mechanics.

A generalized coordinate is called *cyclic*, or "ignorable", if it does not appear explicitly in the Lagrangian. It follows immediately from the corresponding Lagrange's equation that,

$$q_k \text{ is cyclical} \quad \Rightarrow \quad \frac{\partial L}{\partial q_k} = \dot{p}_k = 0. \tag{2.82}$$

In words, *the momentum associated with a generalized cyclic coordinate is constant.*

The connection with symmetries is clear: If the Lagrangian and, consequently, the equations of motion, are indifferent to a change of values of a generalized coordinate (that, after all, does not appear in L), this feature reflects the symmetry of the physical system regarding that change. A constant of motion thus obtained is also called *first integral of motion*. Indeed, the resulting equation

$$p_k = \frac{\partial L}{\partial \dot{q}_k} = \text{constant} \tag{2.83}$$

will be a first-order differential equation in q_k. Often the solution of this equation by *quadrature* (that is, in terms of definite integrals of known functions) will suffice to yield $q_k(t)$. Even so, we still have to solve the $(n-1)$-dimensional problem having second-order equations. Transformations that turn some coordinates to cyclic will play an important role in Hamiltonian dynamics and the reason for a cyclic coordinate also be called ignorable will be clarified then.

Example 2.6 A system of N interacting particles has its Lagrangian given in terms of its CM (X, Y, Z) and relative $\{q_1, \ldots, q_n\}$ coordinates. If the system is isolated, there are no potential energy terms dependent on X, Y, or Z, so that L has the form

$$L = \frac{M}{2}(\dot{X}^2 + \dot{Y}^2 + \dot{Z}^2) + L'(q, \dot{q}). \tag{2.84}$$

The generalized coordinates X, Y eZ are cyclic, thus the associated momenta P_x, P_y e P_z are constants of motion. In this way we obtain linear momentum conservation of an isolated system.

$$\frac{d}{dt}\vec{P} = 0. \tag{2.85}$$

In turn, the cyclic character of the CM coordinates stems from the global translational symmetry of the system, connected to the homogeneity of space.

Example 2.7 Similarly, spatial isotropy indicates that rigid global rotation of an isolated system does not affect its dynamics. This means that the Lagrangian of the system will not depend on any of the angles that characterize the rotation. A formal demonstration, in this case, requires the use of infinitesimal transformations (because finite rotations are not represented by vectors) and is not directly connected to invariance due to a cyclic coordinate. On the other hand, a logical reasoning might be enough: Consider a system of particles undergoing successive infinitesimal rigid rotations around the Cartesian X_i axes. Being the system isolated, there will be no potential energy involved with them, so that, for each direction, $L_i = \dfrac{I_i}{2}\dot{\theta}_i^2$ where θ_i and I_i are, respectively the infinitesimal angles of rigid rotations and the corresponding moments of inertia relative to axis i. In all these Lagrangians the angles are cyclic, so that, according to Eq. (2.83), the corresponding generalized momenta are conserved. But the generalized momenta corresponding to an angular generalized coordinate is an angular momentum, as we already know. So, the conserved quantities are the X, Y, Z components of the total vector angular momentum, which is thus conserved.

Example 2.8 The problem of a particle subject to a central potential $V(r)$ brings out a so far unforeseen situation.

In spherical coordinates, see Fig. 2.10, the Lagrangian of the particle is written as

$$L = \frac{m}{2}(\dot{r}^2 + r^2\dot{\theta}^2 + r^2\dot{\varphi}^2 sen^2\theta) - V(r). \tag{2.86}$$

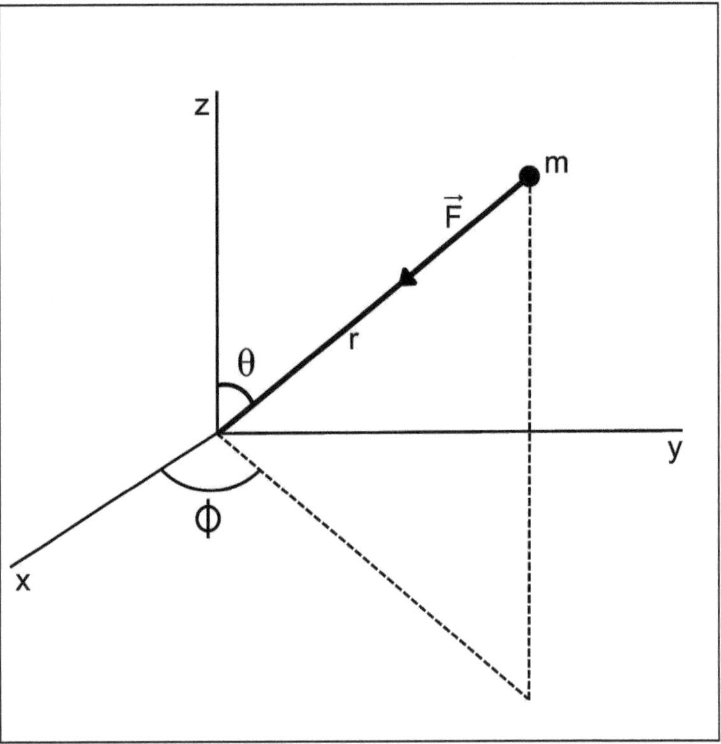

Fig. 2.10 Particle in a central field, spherical coordinates

Exercise Check this form for L.

Despite the lack of real constraints, the movement takes place in a plane. That is, we would supposedly have $n = 3 - 0 = 3$ degrees of freedom, but only two generalized coordinates are needed. In vector terms this plane motion is trivially proven: Being the force in the direction of $\vec{r}$, its torque relative to the origin is zero so, according to Newtonian mechanics, the vector angular momentum $\vec{l}$ is conserved. But, since $\vec{l} = \vec{r} \times \vec{p}$, the velocity and linear momentum vectors must be in a plane perpendicular to $\vec{l}$.

Let us see how this result can fit the structure of Lagrangian mechanics displayed so far. The coordinate φ is cyclic, so that its associated momentum,

$$p_\varphi = \frac{\partial L}{\partial \dot{\varphi}} = mr^2\dot{\varphi}sen^2\theta, \tag{2.87}$$

is a constant of motion. But this result seems odd because the angular momentum of the particle rotating around the Z axis should be simply $mr^2\dot{\varphi}$, which is satisfied only if

$$\theta - \frac{\pi}{2} = 0. \tag{2.88}$$

This is indeed equivalent, at least from a formal point of view, to an equation of holonomic constraint, reducing the dimensionality of the problem to two. The Lagrangian becomes that of Eq. (2.35)

$$L = \frac{m}{2}(\dot{r}^2 + r^2\dot{\varphi}^2) - V(r), \tag{2.89}$$

clearly in plane polar coordinates. The coordinate θ is a redundant independent coordinate. The coordinate φ keeps being cyclic, but now p_φ refers to the modulus of the vector total angular momentum.

Since there is no real constraint in the problem, we could have made an unhappy choice, treating the problem with coordinates x, y, z. In this case, since $r = \sqrt{x^2 + y^2 + z^2}$, there would be no cyclic coordinate and the simplification arising from the existence of a constant of motion (or first integral) would not exist. We might establish a heuristic rule that *the existence of a cyclic coordinate can make redundant another coordinate* provided the last does not appear in the potential. For this, the set of generalized coordinates to be chosen should *make the symmetries of the potential as explicit as possible*. In the present example, the set q should contain the symmetry coordinate r. The next Example explores further this feature.

Example 2.9 Let us consider a constraint-free particle in the gravitational field near Earth's surface, with $\vec{g}$ in direction $-z$. In principle it has 3 degrees of freedom, so that x, y, z can be the generalized coordinates. The Lagrangian is

$$L = \frac{m}{2}(\dot{x}^2 + \dot{y}^2 + \dot{z}^2) + mgz. \tag{2.90}$$

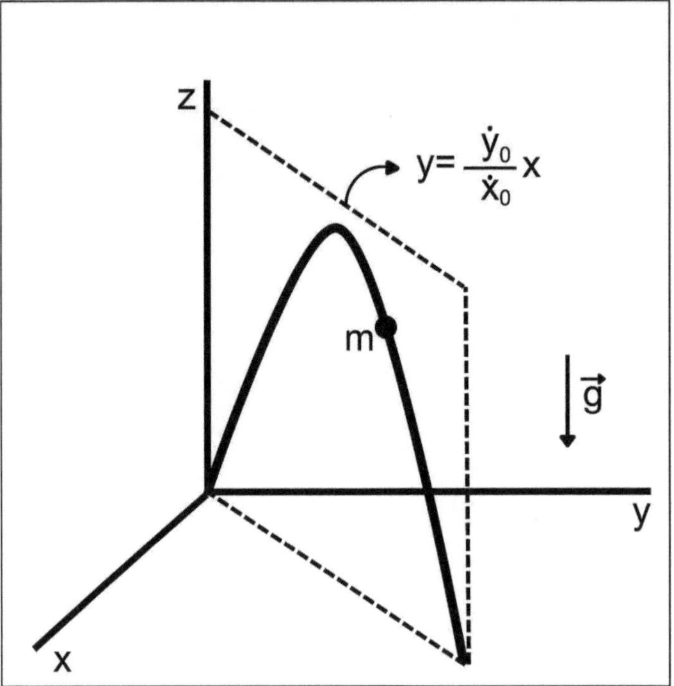

Fig. 2.11 Projectile motion is 2D

Here again we can see that, due to symmetry of the problem, coordinates x and y are cyclic, so the motion also takes place in a plane, see Fig. 2.11. The generalized momenta

$$p_x = \frac{\partial L}{\partial \dot{x}} = m\dot{x} \quad e \quad p_x = \frac{\partial L}{\partial \dot{y}} = m\dot{y} \tag{2.91}$$

are both constants. Calling $\dot{x} = \dot{x}_0$ e $\dot{y} = \dot{y}_0$ and integrating, we get $x = \dot{x}_0 t$ and $y = \dot{y}_0 t$. Eliminating t from these equations, we obtain a linear relationship between them,

$$y = \frac{\dot{y}_0}{\dot{x}_0}x, \tag{2.92}$$

which makes one of them redundant. Again, Eq. (2.92) has the mathematical status of a constraint equation. The axes can then be rotated so that the plane of motion coincides with the xz plane, and only x and z are used as the generalized coordinates for the problem.

2.8 Nöther's Theorem

Some finite transformations of coordinates and time can keep the action integral invariant, as in the following trivial example.

Example 2.10 Consider the 1D harmonic oscillator with Lagrangian $L' = \dfrac{m}{2}\dot{x}^2 - \dfrac{k}{2}(x+\lambda)^2$. This Lagrangian was obtained from the standard form L for an oscillator subjected to the finite translation $x' = x + \lambda$, $t' = t$, with a constant λ. But when we develop the potential energy term, we obtain $L' = L - k\lambda x - \frac{k}{2}\lambda^2$, that is, L' and L are equivalent (check it!), which implies $S' = S$.

However, in order to develop a universal formal treatment of the relationship between transformation symmetries and conservation laws, we must consider infinitesimal transformations. Recall that transformations involving rotations are valid only in the infinitesimal limit. This treatment takes place through Nöther's theorem [4]. Consider the infinitesimal transformations defined by,

$$t' = t + \epsilon X(q, t), \qquad q'_k = q_k + \epsilon Y_k(q, t), \tag{2.93}$$

where X and Y_k are known functions of q and t, and ϵ is an arbitrary infinitesimal parameter. These transformations are used to formalize symmetry operations, granted by demanding the invariance of the action S. In fact, the transformed Lagrangian L' can change, as we have already seen, but it must lead to the same equations of motion. This is guaranteed by requiring that $\Delta S = S' - S = 0$ (not to be confused with the functional variation δS) when the Lagrangian changes from L to L'. This condition, under transformation (2.93), will lead to a conservation principle connected to the involved symmetry.

Under Eq. (2.93), the transformed action S' becomes,

$$S' = \int_{t'_1}^{t'_2} L' dt' = \int_{t'_1}^{t'_2} L\left(q', \frac{dq'}{dt'}, t'\right) dt'. \tag{2.94}$$

where $q'_k = q'_k(t')$ and so on. In order to compare S' with S, they must be expressed as functions of the same time variable, say t. For this, it is helpful to use, from Eqs. (2.93), that $\dfrac{dt'}{dt} = 1 + \epsilon \dot{X}$ so that $\dfrac{dt}{dt'} \simeq 1 - \epsilon \dot{X}$, where the term in ϵ^2 is neglected, as usual. In the same way, $\dfrac{dq'_k}{dt'} = \dfrac{dq'_k}{dt}\dfrac{dt}{dt'} = (\dot{q}_k + \epsilon \dot{Y}_k)(1 - \epsilon \dot{X}) \simeq \dot{q}_k + \epsilon(\dot{Y}_k - \dot{q}_k \dot{X})$. In consequence,

$$\Delta S = \int_{t_1+\epsilon X(t_1)}^{t_2+\epsilon X(t_2)} L[q+\epsilon Y, \dot{q}+\epsilon(\dot{Y}-\dot{q}\dot{X}), t+\epsilon X](1+\epsilon \dot{X})dt - \int_{t_1}^{t_2} L(q,\dot{q},t)dt = 0.$$
$$(2.95)$$

Now, concerning the integration limits, it is easily verified, using binomial expansions, that there will be a part of ΔS involving only an integral between t_1 and t_2 added to another part multiplying the parameter ϵ.

Exercise Verify this statement with simple functions.

Being ϵ arbitrary, these parts must vanish independently, so that we can write

$$\Delta S = \int_{t_1}^{t_2} [L(q,\dot{q},t) + \Delta L](1 + \epsilon \dot{X})dt - \int_{t_1}^{t_2} L(q,\dot{q},t)dt = 0. \qquad (2.96)$$

Explicitly, $\Delta L = \epsilon \sum_{k=1}^{n} [\frac{\partial L}{\partial q_k}Y_k + \frac{\partial L}{\partial \dot{q}_k}(\dot{Y}_k - \dot{q}_k \dot{X}) + \frac{\partial L}{\partial t}X]$, and as we develop the algebra and disregard terms in ϵ^2, we get

$$\Delta S = \epsilon \int_{t_1}^{t_2} \{[\sum_{k=1}^{n}[\frac{\partial L}{\partial q_k}Y_k + \frac{\partial L}{\partial \dot{q}_k}(\dot{Y}_k - \dot{q}_k \dot{X})] + L\dot{X} + \frac{\partial L}{\partial t}X\}dt = 0. \qquad (2.97)$$

Since ϵ and the integration interval are arbitrary, this identity implies the integrand to be zero.

This is the appropriate point to introduce *Jacobi's integral h* as,

$$h = \sum_{k=1}^{n} \frac{\partial L}{\partial \dot{q}_k}\dot{q}_k - L. \qquad (2.98)$$

Inserting it into the vanishing integrand of Eq. (2.97) we obtain (check it!),

$$\frac{d}{dt}[\sum_{k=1}^{n} \frac{\partial L}{\partial \dot{q}_k}Y_k - hX] = 0. \qquad (2.99)$$

As we replace in this equation the expression of h above, after some manipulation, we find, in conclusion, that demanding S to be stationary under the symmetry transformations (2.93), the associated constant of motion is

$$C = \sum_{k=1}^{n} \frac{\partial L}{\partial \dot{q}_k}(\dot{q}_k X - Y_k) - LX, \qquad \dot{C} = 0. \qquad (2.100)$$

Exercise Choose X and Y_k in order to explain the constant character of the momentum associated to a cyclic coordinate q_l. Discuss the case of a rigid translation of a system.

2.8.1 *Back to Conservation of the Angular Momentum*

In the scope of lagrangian mechanics, the conservation of the angular momentum of an isolated system of particles is done within Nöther's principle. It is easier to do it if we resort to vector notation, but only for economy. The infinitesimal rigid rotation is represented by $\delta \vec{\theta} = \epsilon \mathbf{n}$, where $\mathbf{n}$ is an arbitrary unit vector, and we take $X = 0$ and $\delta \vec{r}_\alpha = \delta \vec{\theta} \times \vec{r}_\alpha = \epsilon \mathbf{n} \times \vec{r}_\alpha = \epsilon Y_\alpha$ for all particles. Thus, since a rigid rotation is a symmetry operation for an isolated system, there will be a constant of motion,

$$C = -\sum_{\alpha=1}^{N} \frac{\partial L}{\partial \dot{\vec{r}}_\alpha} \cdot \mathbf{n} \times \vec{r}_\alpha = -\sum_{\alpha=1}^{N} \vec{p}_\alpha \cdot \mathbf{n} \times \vec{r}_\alpha. \tag{2.101}$$

Now, using the permutation property of the triple product, we obtain $C = -\mathbf{n} \times \sum_{\alpha=1}^{N} \vec{r}_\alpha \times \vec{p}_\alpha = -\mathbf{n} \times \vec{L}$, which, being $\mathbf{n}$ arbitrary in direction, implies the conservation of the total angular moment $\vec{L}$.

2.9 Energy—Jacobi's *h* Integral

A conservation principle of special significance, the conservation of mechanical energy, is connected to the homogeneity of time in classical mechanics. As time is not a generalized coordinate (at least in the present formulation), we cannot deal with this issue in terms of a cyclic coordinate. But, using Nöther's theorem, we arrive at a constant of motion of singular importance. Initially, we must realize that the infinitesimal transformation (2.93) with $X = 1$ and $Y_k = 0$ corresponds to a infinitesimal time translation $t' = t + \epsilon$. These values taken into Eq. (2.97) yield $\frac{\partial L}{\partial t} = 0$. *Time homogeneity* is then connected to the explicit independence of the Lagrangian on time,

$$t' = t + \epsilon \quad \Leftrightarrow \quad \frac{\partial L}{\partial t} = 0. \tag{2.102}$$

In turn, taking the same values of X and Y_k in Eq. (2.100), we produce the constant of motion, which results as the quantity h defined in (2.98), the Jacobi integral,

$$C = h = \sum_{k=1}^{n} p_k \dot{q}_k - L.$$

(2.103)

On the other hand, taking the total derivative of h in Eq. (2.98) and using Lagrange's equations, we easily find

$$\frac{dh}{dt} = -\frac{\partial L}{\partial t}.$$

(2.104)

This last result establishes that, *if the Lagrangian does not depend explicitly on time, h is conserved.* This result should be compared with the previous conservation principles. Despite t being not a generalized coordinate, its explicit absence in L is still connected (as if t was a cyclic coordinate) to the conservation of a dynamical quantity. It remains to give h a physical interpretation.

Exercise Derive the result expressed in Eq. (2.104).

2.9.1　Energy Conservation

In order to connect h with a known physical quantity, let us consider an isolated system of N particles, identified with the index α. Symbolically, the transformation equations are

$$x_{\alpha i} = x_{\alpha i}(q_j, t), \quad i = 1, 2, 3$$

(2.105)

and, according to Eq. (2.17), the generalized velocities are,

$$\dot{x}_{\alpha i} = \sum_{j=1}^{n} \frac{\partial x_{\alpha i}}{\partial q_j} \dot{q}_j + \frac{\partial x_{\alpha i}}{\partial t},$$

(2.106)

The kinetic energy of the system can then be written as

$$T = \sum_{\alpha}^{N} \frac{m_{\alpha}}{2} \sum_{i=1}^{3} \dot{x}_{\alpha i}^{2} = \sum_{j,k=1}^{n} a_{jk} \dot{q}_j \dot{q}_k + \sum_{j=1}^{n} b_j \dot{q}_j + c,$$

(2.107)

in which,

$$a_{jk}(q_j, t) = \sum_{\alpha=1}^{N} \sum_{i=1}^{3} \frac{m_\alpha}{2} \frac{\partial x_{\alpha i}}{\partial q_j} \frac{\partial x_{\alpha i}}{\partial q_k}, \tag{2.108}$$

$$b_j(q_j, t) = \sum_{\alpha=1}^{N} \sum_{i=1}^{3} m_\alpha \frac{\partial x_{\alpha i}}{\partial q_j} \frac{\partial x_{\alpha i}}{\partial t},$$

$$c(q_j, t) = \sum_{\alpha=1}^{N} \sum_{i=1}^{3} \frac{m_\alpha}{2} \left(\frac{\partial x_{\alpha i}}{\partial t}\right)^2.$$

Note that the first term is quadratic in the velocities and time-independent. The second is linear in the velocities and the third is velocity-independent, and both time-dependent.

It follows that,

$$\frac{\partial T}{\partial \dot{q}_l} = \sum_{k=1}^{n} a_{lk} \dot{q}_k + \sum_{j=1}^{n} a_{jl} \dot{q}_j + b_l. \tag{2.109}$$

Multiplying by $\dot{q}_l$ and summing over l, we get

$$\sum_{l=1}^{n} \frac{\partial T}{\partial \dot{q}_l} \dot{q}_l = \sum_{l,k=1}^{n} a_{lk} \dot{q}_l \dot{q}_k + \sum_{j,l=1}^{n} a_{jl} \dot{q}_j \dot{q}_l + \sum_{l=1}^{n} b_l \dot{q}_l \equiv 2T^{qd} + \sum_{l} b_l \dot{q}_l, \tag{2.110}$$

where T^{qd} represents the part of the kinetic energy that depends quadratically of the generalized velocities. This is true because the two sums at the right-hand-side are fully equivalent to the first term at the right-hand-side of Eq. (2.107).

We can now evaluate Jacobi's integral, Eq. (2.103) as

$$h = \sum_{l=1}^{n} \frac{\partial T}{\partial \dot{q}_l} \dot{q}_l - L = 2T^{qd} + \sum_{l=1}^{n} b_l \dot{q}_l - T^{qd} - \sum_{l=1}^{n} b_l \dot{q}_l - c + V(q), \tag{2.111}$$

where we have already assumed that the potential energy does not depend on the generalized velocities, so that $\frac{\partial L}{\partial \dot{q}_k} = \frac{\partial T}{\partial \dot{q}_k}$ (**condition 1**). An important observation is that the terms which depend linearly in the velocities, $\sum_{l=1}^{n} b_l \dot{q}_l$, always cancel in h. In that case, we have, by the expression above,

$$h = T^{qd} + V(q) - c(q, t). \tag{2.112}$$

This result suggests a relationship between the constant h and the total energy of the system, $E = T + V$. To deal with the term $c(q, t)$, a second condition (**condition 2**) normally imposed is that the transformation equations do not depend explicitly on time. This condition implies $\dfrac{\partial x_{\alpha i}}{\partial t} = 0$ in Eqs. (2.108), so that we have both $c(q, t) = 0$ and $b_j(q_j, t) = 0$, the former being necessary for h to be equal to $T^{qd} + V$. It is also common to admit, from the beginning, that the kinetic energy has exclusively the quadratic form T^{qd}, so that $b_j(q_j, t)$ and $c(q, t)$ are assumed null in Eqs. (2.108). In any case, the possibility of moving constraints is excluded.

The conclusion is powerful: From Eq. (2.104), *if* $\dfrac{\partial L}{\partial t} = 0$, *h is a conserved quantity*. We can call it **condition 3**. In turn, in compliance with conditions 1 and 2 above,

$$h = E, \tag{2.113}$$

the energy E being a constant of motion.

The three conditions are independent, so we can have situations with different results. (i) In the very important particular case where all conditions are satisfied, we conclude for the conservation of the mechanical energy E. Time homogeneity leads us, then, to the conservation of a quantity, the Jacobi integral h, which, under specific conditions, is equal to the energy of the system. (ii) However, even if E is not conserved, h can be, as long as $\dfrac{\partial L}{\partial t}$ is null. This is the case of the Lagrangian (2.72), for which condition 3 is satisfied, but in which the energy of the bead is clearly increasing.

Exercise Prove that h is constant for the problem leading to Eq. (2.72). Hint: Calculate $\dfrac{dh}{dt}$ and use the equation of motion (2.73).

(iii) Another different situation happens in cases where $V = V(q, t)$, so that we can have $h = E$, under conditions 1 and 2, but neither of them being conserved. As an example, an electric charge e subjected to an oscillating electric field $\varepsilon \cos(\omega t)$, whose 1D Lagrangian is $L = \dfrac{m}{2}\dot{x}^2 - ex\varepsilon\cos(\omega t)$. h is not conserved since L is explicitly time-dependent, but $h = E$ based on conditions 1 and 2. (iv) Another singular situation, in which condition 2 does not met, refers to the Galilean invariance of Eq. (2.113), and is explored in the next section.

The disappearance of the linear terms in the generalized velocities, in the expression of h, Eq. (2.112), is fundamental for the relevance of this quantity. As we have already seen, terms like these either cancel each other in Lagrange's equations or even have a null effect on the equations of motion, due to the equivalence of different Lagrangians, but do not necessarily cancel each other in the expression of total energy E.

Constants of motion are always of special importance. We'll see further that, in any case, Jacobi's integral h has a position of great relevance in further developments of mechanics.

2.9.2 Energy and Galilean Relativity

As we know from basic mechanics, if an isolated system has uniform global translation, its Lagrangian has the form

$$L = \frac{M}{2}\vec{u}^2 + L_{int},\tag{2.114}$$

where M is its total mass and $\vec{u} = \vec{V}_{CM}$ the translational speed of its center of mass (see also Sect. 1.9.2).

Exercise Use relative and CM coordinates to prove Eq. (2.114).

Its internal Lagrangian L_{int} is not affected by the global movement, being unable to change its total energy E. But what about h, is it also constant? Is it equal to E? The questions are justified, since, as seen in Example 2.5, there is a temporal dependence of the transformation equations in a case like this and, therefore, by condition 2, we should have $h \neq E$.

As a matter of fact, condition 2, in addition to having an exception as we shall see, usually induces mistakes concerning the very existence of Jacobi's integral h [5]. Particular situations involving condition (2) have to do with Galileo's principle of relativity and with moving constraints [6]. Considering first the relativity principle, let us return to the matter of global translation of a system. The transformation equations depend linearly on time,

$$x_{\alpha i}(q,t) = u_i t + g_{\alpha i}(q),\tag{2.115}$$

where $g_{\alpha i}(q)$ is a generic function, but now the factor c of Eqs. (2.108) becomes constant and equal to translational kinetic energy of the system,

$$c(q,t) = \sum_{\alpha=1}^{N}\sum_{i=1}^{3}\frac{1}{2}m_\alpha\left(\frac{\partial x_{\alpha i}}{\partial t}\right)^2 = \frac{1}{2}Mu^2,\tag{2.116}$$

where $M = \sum_{\alpha} m_\alpha$ is the total mass. This result, taken into Eq. (2.112), finally produces,

$$h = T^{qd} + V(q) - \frac{1}{2}Mu^2, \quad \Rightarrow \quad E = h + \frac{1}{2}Mu^2.\tag{2.117}$$

Being u constant, $\dfrac{\partial L}{\partial t} = 0$ because the transformation equations do not introduce a t dependence on the kinetic energy T, then h is constant of motion. Despite the transformation equations (2.115) depend explicitly on time, relation (2.117) above is completely equivalent to (2.113), since it implies that E is constant. E and h differ by a constant equal to the translational kinetic energy of the system. Both are invariant under Galileo transformation, where h here represents the internal energy of the system. Above all, the conservation of E as a consequence of conservation of h does not demand that $c(q, t)$ is zero, but simply a constant. In conclusion, equation $h = E = constant$ is Galileo invariant, condition 2 is too strong and must be used with care.

2.9.3 Rheonomic Systems and Constants of Motion

It is instructive to reconsider the examples in Sect. 2.6 from the present perspective. In the case of the bead limited to moving along a rod that rotates with constant angular velocity, Example 2.4, we must first realize that, although the virtual work of the force of constraint on the bead is null, the actual work is not. The rod does positive work on it and, as a result, E is not constant. On the other hand, as the Lagrangian (2.72) does not depend explicitly on t, or $\dfrac{\partial L}{\partial T} = 0$, h is constant. It's not hard to see why we have formally $h \neq E$, in this case. In fact, the transformation equations (2.71) depend explicitly on time in a non-linear way. The term $\frac{1}{2}m\omega^2 r^2$ of Lagrangian (2.72) has origin precisely in this temporal dependence. Indeed, referring to definition of $c(q, t)$ in Eq. (2.108),

$$c(q, t) = \frac{m}{2}\left[\left(\frac{\partial x}{\partial t}\right)^2 + \left(\frac{\partial y}{\partial t}\right)^2\right] = \frac{m}{2}\omega^2 r^2. \tag{2.118}$$

So $c(q, t)$ is not constant and the second condition, which prevents us from having $h = E$, here works flawlessly.

It is a common mistake to consider that the explicit time-dependence of the transformation equations or, equivalently, the existence of rheonomic (time-dependent) constraints, prevents the conservation of the Jacobi integral h [5]. On the other hand, this dependency can really prevent the equality $h = E$ from being fulfilled, now due to factors $b_j(q_j, t)$ in Eqs. (2.108). We have already seen that these factors cancel each other in the expression of h, but they do not necessarily cancel each other in the expression of E.

To exemplify this point, let us now return to the pendulum in a car of Fig. 2.7 but with $a = 0$, Example 2.5. The conclusions about h and E will be analogous to the bead case in Example 2.4, that is, h is constant because $\dfrac{\partial L}{\partial t} = 0$, while $E \neq h$, since the first transformation Eq. (2.65) depends on t. On the other hand, the interpretation of the results is more engaging, not least because the pendulum support is not accelerated. Since the same equations of motion are generated in the two references (fixed or mobile support), there is a tendency for the student to resort to the principle of relativity to conclude, erroneously, that $h = E$. In reality, E cannot be conserved, neither for the pendulum nor for the system as a whole, which is not isolated since an external force is required to maintain uniform the movement of the constraint-pendulum system. That is, the constraint does work on the pendulum. Analysing the problem in more detail, from Lagrangian (2.75), we obtain

$$p_\theta = \frac{\partial L}{\partial \dot\theta} = ml^2\dot\theta + mul\cos\theta \tag{2.119}$$

and

$$h = p_\theta\dot\theta - L = \frac{1}{2}ml^2\dot\theta^2 - mgl\,sen\theta - \frac{1}{2}mu^2. \tag{2.120}$$

Again, h is a constant of motion, while the energy,

$$E = \frac{1}{2}mu^2 + \frac{1}{2}ml^2\dot\theta^2 + mul\dot\theta\cos\theta - mgl\,sen\theta, \tag{2.121}$$

is not constant. There is no inconsistency between this result and the previous ones. The dependency on t of the transformation equations is still linear since $a = 0$, so the expression (2.112) for h remains valid, as the terms $\pm mul\dot\theta\cos\theta$ cancel out. Besides that, note that E refers only to the pendulum, not to the whole system car plus pendulum. On the other hand, although $c(q, t)$ remains constant, now,

$$E \neq T^{qd} + V, \tag{2.122}$$

since the corresponding positive sign term (coming from $b_j(q_j, t)$) of Eq. (2.108) remains in the expression of E. In qualitative terms, E captures the oscillatory transfer of energy from the constraint to the pendulum, being constant only in time average. The same term is also present in L, Eq. (2.75), but without consequences on the equations of motion, as we have seen, in a clear manifestation of the symmetry of physical laws.

In the particular case of uniform motion of the constraint, the factors $b_j(q_j, t)$ cancel out both in h and in E, but they cancel only in h in the general cases of accelerated constraints. These features advance a fundamental importance of this quantity, h, which will play a role analogous to L as generator of the equations of motion in Hamiltonian mechanics.

2.10 The General Motion of a Rigid Body

Lagrangian theory of general rigid-body motion also shows some advantages over its Newtonian counterpart, namely, the Euler equations approach. The aim of this section is to compare the two approaches.

A rigid-body can be modelled as a system of $N(> 2)$ particles linked pairwise by rigid constraints, in such a way that the mutual distances are fixed. However, the corresponding constraint equations are not all effective in reducing the body's degrees of freedom, as we have already discussed in Sect. 2.1.2. The number of degrees of freedom for such a rigid body is easily obtained by an alternative way. Suppose the body moving in the 3D space. If we take a point on the body, it will have three degrees of freedom (relative to a reference frame external to the body). A second point will only have two degrees of freedom, because there will be a constraint equation involving its coordinates to those of the first point. In fact, it will be free to move only in a spherical surface with the first point at is center. Finally, by an identical reasoning, a third point will only have a single degree of freedom; it will only be able to rotate around the axis that goes through the first two. Other points will not generate additional degrees of freedom, that is, a rigid-body has $n = 6$. Of these, three are translational degrees of freedom, usually the Cartesian coordinates of the center of mass (CM) of the body. The other three generalized coordinates, associated with rotation of the rigid-body, are normally the well-known Euler angles (ϕ, θ, ψ), considered as angular coordinates that simplify the description of the movement as much as possible.

Consider the rigid-body rotating with angular speed $\vec{\omega}$ around its instantaneous axis of rotation, as well as translating, see Fig. 2.12.

Consider also an inertial reference frame (LAB) in the laboratory, to refer to the translation of a fixed point Q of the body, and another frame (BODY), attached to

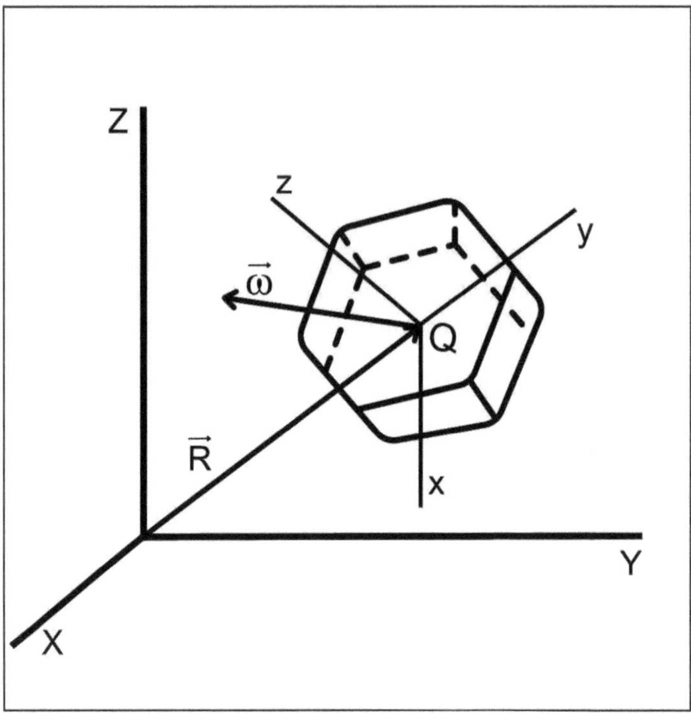

Fig. 2.12 A rigid body and reference frames

the body, with its origin at Q, as in the figure. It is shown in basic texts that, once the position vectors of the rigid body particles are referred to its CM ($Q = $CM), then the kinetic energy of the body, relative to the LAB frame, can be decomposed in the form

$$T = T_{tr} + T_{rot} = \frac{M}{2}(\dot{X}^2 + \dot{Y}^2 + \dot{Z}^2) + T_{rot}. \tag{2.123}$$

where X, Y, Z stand for the CM coordinates and T_{rot} for the rotational energy.

An easier situation happens if there is a point of the body that is fixed relative to the LAB frame. In this case the origin of the two frames can be made to coincide, so that the translation can be ignored. A typical case (the top toy) is discussed in an Example that follows. The possible existence of external torques makes the potential energy to depend on the Eulerian angles.

The potential energy is also separable in translation and rotation,

$$V = V_{tr}(X, Y, Z) + V_{rot}(\phi, \theta, \psi), \tag{2.124}$$

so that the Lagrangian relative to the LAB frame can be partitioned in a translational and a rotational parts,

$$L = L_{tr} + L_{rot} = [\frac{M}{2}(\dot{X}^2 + \dot{Y}^2 + \dot{Z}^2) - V_{tr}(X, Y, Z)] + [T_{rot} - V_{rot}(\phi, \theta, \psi)], \tag{2.125}$$

and will generate two separate Lagrange equations (a trivial property of linear differential equations). Here we will focus on the rotational L_{rot}, since the translational problem, equivalent to that of a particle with mass M, has been already treated. Common examples are: free rotation (no external forces, $V_{rot} = 0$); rotation with a fixed point; and rotation under a single force (e.g., gravity) acting on the CM of the body (no torque, $V_{rot} = 0$). Fortunately, many of problems of interest reduce to one of these cases; for example, a body thrown into the gravitational field, which reduces to the last case. Finally, we must remember again that constraint forces (static friction, for instance) do not "disappear" from the problem, only from the equations of motion. This means that possible torques due to constraints will be automatically ruled out via the choice of generalized coordinates consistent with constraint equations (as in the purely translational case). The cases of planar rotation are examples of decreasing of degrees of freedom, $n = 1$ instead of 3 due to constraints (see Example 2.11 ahead).

The inertia tensor $\overleftrightarrow{I}$ have constant matrix elements in the BODY frame. If it refers to a set of *principal axis* it will be diagonal, which corresponds to the most favourable situation. In this case, the rotational kinetic energy reduces to the simple expression,

$$T_{rot} = \frac{1}{2} \sum_{i=1}^{3} I_i \omega_i^2, \tag{2.126}$$

while the Lagrangian for rotation is

$$L_{rot} = T_{rot} - V_{rot} = \frac{1}{2}\sum_{i=1}^{3} I_i \omega_i^2 - V_{rot}(\phi, \theta, \psi), \qquad (2.127)$$

where the I_i are the diagonal elements of the inertia tensor and the ω_i are components of the angular velocity vector $\vec{\omega}$. This expression takes into account that we refer to an inertial frame of reference, even tough $\overleftrightarrow{I}$, is represented, conveniently, in the body's frame of reference.[4]

We want to use Eulerian angles as our generalized coordinates, but here there is a problem: The kinematics of rigid-bodies is not really similar to that of a particle. Particularly, there is no "vector angular position" such that its time-derivative would yield $\vec{\omega}$. Instead, we know from elementary vector physics that, for a particle in instantaneous rotation around an axis, the angular velocity vector is defined so as to have $\vec{v} = \vec{\omega} \times \vec{r}$. Moreover, there are no trivial transformation equations relating the Cartesian coordinates of the LAB frame to the Eulerian angles but, evidently, it is possible to write directly the *components* of $\vec{\omega}$ in terms of the derivative of the angles, namely $\dot{\phi}, \dot{\theta}, \dot{\psi}$. This is what we are left with and, since ϕ, θ, ψ are defined by successive rotations around the BODY axis, this representation is obtained in elementary textbooks by rotation matrices, leading to

$$\omega_x = \dot{\phi}_x + \dot{\theta}_x + \dot{\psi}_x = \dot{\phi}\,sen\theta\,sen\psi + \dot{\theta}\,cos\psi, \qquad (2.128)$$

$$\omega_y = \dot{\phi}_y + \dot{\theta}_y + \dot{\psi}_y = \dot{\phi}\,sen\theta\,cos\psi - \dot{\theta}\,sen\psi,$$

$$\omega_z = \dot{\phi}_z + \dot{\theta}_z + \dot{\psi}_z = \dot{\phi}\,cos\theta + \dot{\psi}.$$

Once we know the expression for $V(\phi, \theta, \psi)$, we have all the elements needed to set the Lagrangian of the rotation of a rigid-body.

A last previous consideration is that, by the very definitions of the Euler angles, only the axis rotation that defines ψ (the only rotation around a BODY axis, x_3) corresponds to the component of the total torque relative to a principal axis of the BODY frame. Thus, different from particle motion with Cartesian generalized coordinates, only one of the 3-components of vector Euler's equation, $\vec{\tau} = \dot{\vec{L}}$, is identified with one of Lagrange's equation. Here $\vec{\tau}$ and $\vec{L}$ are, respectively, the total torque and the angular momentum[5] of the body.

Hence, for the generalized coordinate ψ,

$$\frac{\partial(T_{rot} - V_{rot})}{\partial \psi} - \frac{d}{dt}\frac{\partial(T_{rot} - V_{rot})}{\partial \dot{\psi}} = 0, \qquad (2.129)$$

---

[4] Because the inertia tensor is diagonal! If the reader has doubts on this point, he/she should consult a basic text on Newtonian mechanics.

[5] Being the angular momentum a vector quantity, there is no problem in keeping its standard symbol L just here.

but since $\dfrac{\partial V_{rot}}{\partial \dot{\psi}} = 0$ and $\tau_z = -\dfrac{\partial V_{rot}}{\partial \psi}$, we arrive to

$$\frac{d}{dt}\frac{\partial T_{rot}}{\partial \dot{\psi}} - \frac{\partial T_{rot}}{\partial \psi} = \tau_z. \tag{2.130}$$

Instead of direct replacement of Eq. (2.126) for T_{rot}, it is more practical to transform Eq. (2.130) to,

$$\sum_{i=1}^{3}[\frac{d}{dt}(\frac{\partial T_{rot}}{\partial \omega_i}\frac{\partial \omega_i}{\partial \dot{\psi}}) - \frac{\partial T_{rot}}{\partial \omega_i}\frac{\partial \omega_i}{\partial \psi}] = \tau_z. \tag{2.131}$$

Now, from Eqs. (2.128), we get

$$\frac{\partial \omega_x}{\partial \psi} = \omega_y, \ \frac{\partial \omega_y}{\partial \psi} = -\omega_x, \ \frac{\partial \omega_z}{\partial \psi} = 0, \ \frac{\partial \omega_x}{\partial \dot{\psi}} = \frac{\partial \omega_y}{\partial \dot{\psi}} = 0, \ \frac{\partial \omega_z}{\partial \dot{\psi}} = 1, \ \frac{\partial T_{rot}}{\partial \omega_i} = I_i\omega_i. \tag{2.132}$$

Taking these results into Eq. (2.131), we come to

$$I_z\dot{\omega}_z - \omega_x\omega_y(I_x - I_y) = \tau_z, \tag{2.133}$$

which is the same as the third Euler's Eq. (2.135) bellow. Lagrange's equations for other generalized coordinates ϕ and θ can also be obtained, but, as we saw, $-\dfrac{\partial V_{Rot}}{\partial \phi}$ or $-\dfrac{\partial V_{rot}}{\partial \theta}$ *are not* the torques τ_x or τ_y. For example, Lagrange's equation for ϕ is,

$$\frac{d}{dt}(I_x\omega_x sen\theta sen\psi + I_y\omega_y sen\theta cos\psi + I_z\omega_z cos\theta) = -\frac{\partial V_{rot}}{\partial \phi}, \tag{2.134}$$

where $-\dfrac{\partial V_{rot}}{\partial \phi}$ is the torque related to the third axis of the LAB frame.

Exercise Derive Lagrange's equation for θ, recalling that the torque $-\dfrac{\partial V_{rot}}{\partial \theta}$ will be relative to the line of nodes (see Fig. 2.14 ahead).

The set of three Lagrange's equations is sufficient for the complete solution of the problem. On the other hand, back to the vector equations of motion, identifying one of the principal axis with x_3 is fully arbitrary, so that, in Eq. (2.133) above, the indexes can be exchanged, leading to the other components of $\vec{\tau} = \dot{\vec{L}}$,

$$I_x\dot{\omega}_x - \omega_y\omega_z(I_y - I_z) = \tau_x, \tag{2.135}$$

$$I_y\dot{\omega}_y - \omega_z\omega_x(I_z - I_x) = \tau_y,$$

$$I_z\dot{\omega}_z - \omega_x\omega_y(I_x - I_y) = \tau_z.$$

These are the so-called Euler's equations in components, equally sufficient to completely solve the problem of rotation of a rigid-body, written in an elegant form. They can also be deducted from the vector relation $\overrightarrow{\tau} = \dot{\overrightarrow{L}}$ with $\overrightarrow{L} = \overleftrightarrow{I}\,\overrightarrow{\omega}$. Just the last corresponds to a Lagrange equation. The choice between using Euler's equations or Lagrange's equations depends on the problem under study. Particularly, the Lagrangian method becomes much convenient in the presence of cyclic angle coordinates, as we will see.

Before that, let us see how to apply Lagrange's equations to a simple planar motion of a body.

Example 2.11 A homogeneous disc of radius R and moment of inertia $I = \dfrac{m}{2}R^2$ can rotate without friction around its axis. A small mass m, with negligible effect on the moment of inertia of the disc, is fixed to a point in the edge of the disk, according to Fig. 2.13, causing a gravitational torque upon it. What will be the movement of the disk?

The choice of the symbol ψ for the angle, in Fig. 2.13, is no coincidence. As the axis is fixed, there is only one rotational degree of freedom. Identifying this axis

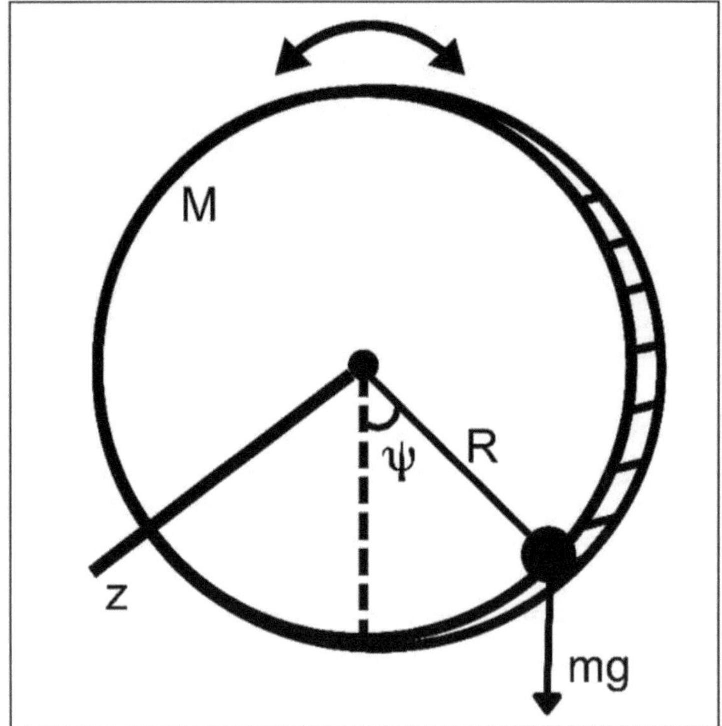

Fig. 2.13 Disk in rolling around a fixed axis (z) subject to a torque

with the z axis fixed in the body, the Lagrange equation becomes Eq. (2.133) with $\omega_x = \omega_y = 0$, $I_z = I$ and $\omega_z = \dot{\psi}$. We then easily get the equation of motion,

$$\ddot{\psi} - \frac{2mg}{MR} sen\psi = 0. \tag{2.136}$$

This is a non-linear equation that must be resolved by a numerical or perturbation method. But for small oscillations, we have $sen\psi \simeq \psi$ and get a linear oscillator equation $\ddot{\psi} - \frac{2mg}{MR}\psi = 0$, which means that the body will oscillate around $\psi = 0$ with frequency $\sqrt{\frac{2mg}{MR}}$.

Equation (2.133) coincides with Euler's third equation and there is no advantage of using Lagrange's method in this situation. As said, however, the Lagrangian method is more helpful in cases having cyclic angular generalized coordinates. The next Example is illustrative of this feature.

Example 2.12 The case of the toy top motion (intimately related to the important gyroscopic movement) is much more conveniently treated by the Lagrangian mechanics. A simplification for this problem is that the top moves having a fixed point (due to constraint forces) in the inertial frame, so that the origin of the principal axes system is conveniently the same, see Fig. 2.14.

Taking z as the axis of symmetry, it is easy to see, from Fig. 2.14, that the gravitational torque is directed along the line of nodes, not the direction of one of the x or y axis. On the other hand, this torque component is $-\dfrac{\partial V_{rot}}{\partial \theta}$ (check it!).

Since $I_x = I_y$, the kinetic energy is

$$T = \frac{I_x}{2}(\omega_x^2 + \omega_y^2) + \frac{I_z}{2}\omega_z^2, \tag{2.137}$$

which can be written in terms of Euler's angles with the aid of Eqs. (2.128). The Lagrangian becomes,

$$L = \frac{I_x}{2}(\dot{\theta}^2 + \dot{\phi}^2 sen^2\theta) + \frac{I_z}{2}(\dot{\psi} + \dot{\phi}cos\theta)^2 - Mglcos\theta. \tag{2.138}$$

We realize immediately that the coordinates ϕ and ψ are cyclic, which leads to the conservation of their associated angular momenta. The momenta p_ϕ and p_ψ are first integrals with constant values,

$$p_\psi = \frac{\partial L}{\partial \dot{\psi}} = I_z(\dot{\psi} + \dot{\phi}cos\theta) = I_z\omega_z, \tag{2.139}$$

$$p_\phi = \frac{\partial L}{\partial \dot{\phi}} = (I_x sen^2\theta + I_z cos^2\theta)\dot{\phi} + I_3\dot{\psi}cos\theta. \tag{2.140}$$

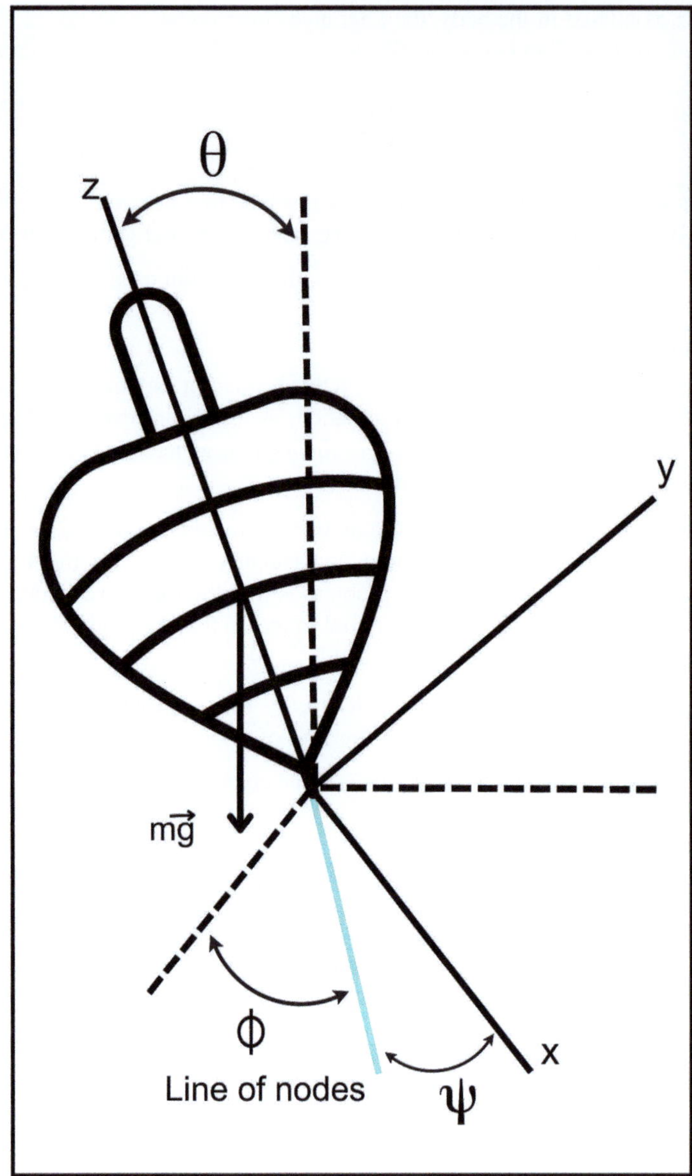

Fig. 2.14 The top toy motion

These results, together with the expression of the constant energy of the top,

$$E = T + V = \frac{I_x}{2}(\dot{\theta}^2 + \dot{\phi}^2 sen^2\theta) + \frac{I_z}{2}\omega_z^2 + Mglcos\theta, \qquad (2.141)$$

makes its movement completely determined by quadrature, which can yield $\phi(t), \theta(t), \psi(t)$, our final solutions. It seems impossible not to get amazed with the elegance of this solution.

2.10.1 Combined Translation and Rotation

The choice of generalized coordinates, as well as the solution of the problem for combined translation and rotation, are only trivial in simplified problems, such as those in which the direction of the rotation axis is fixed, in the absence of slipping. On the other hand, the gain in addressing more complicated problems is not justified here.

Example 2.13 Consider a disc that rolls without slipping along a inclined plane, as in Fig. 2.2. In this simple Example, we must observe that: (1) The torque is produced by static friction, a constraint force, and is automatically taken into consideration with the choice of the lone generalized coordinate ψ. (2) The Lagrangian is separable in a translational and another rotational part, which are, however, linked by the constraint equation (2.12). The problem can thus be put simply in terms of either x or ψ.

2.11 Final Considerations About Lagrangian Mechanics

Obtaining Lagrange's equations from Newtonian mechanics limits us to assume $L = T - V$ and to resort to the principle of virtual work, a not very elegant device to eliminate constraint forces. The consequent reasoning defects propagate from a poorer view of the theory, which would simply be a method of solving problems involving conservative forces.

I hope I have conveyed to the reader a much wider vision of Lagrangian mechanics, basically that all mechanical system should have a Lagrangian function associated with it, so that its equations of motion would be obtained by the application of Hamilton's principle. This point of view is much more consistent with the inductive way in which physical theories are built and, at the same time, combines with initiatives to obtain increasingly comprehensive theories. This is done through the assumption that Hamilton's principle has a superior status in nature.

The reader should have realized the radical change of paradigm: *If the Lagrangian is not previously determined, but Hamilton's principle is assumed to be valid for some form of it, this assumption opens the possibility of going beyond classical mechanics of conservative systems, or even the mechanics of particles and bodies.* The choices of *gauge* potentials and the relativistic Lagrangian formulation are impressive examples of this paradigm. Moreover, a Lagrangian theory of fields, the subject of Chap. 4, could never have been built without this intellectual progress.

Chapter 3
Hamiltonian Mechanics

Abstract This chapter is also equivalent to an almost complete introduction to standard Hamiltonian mechanics, with some new detailing and derivations. How Hamilton's equations really work and, particularly, the difficult concept of adiabatic invariance, are topics treated in a more didactical way. An example of using action-angle variables in research is given.

Keywords Canonical pairs · Phase space · Hamilton's equations in phase space · Canonical transformations · Hamilton-Jacobi theory · Adiabatic invariants · Connection to quantum mechanics

So far, we have emphasized the role of the Lagrangian function in mechanics, under the background of Hamilton's principle. Now, supported by the same principle, we change the paradigm to put emphasis on special pairs of variables, called *canonical variables*, together with the equations they obey, Hamilton's equations.

Hamiltonian mechanics, which we'll be dealt with in this chapter, is not necessarily competitive for the solution of didactical problems in mechanics, although being very useful in numerical simulations of many-particle systems, because it deals with first-order equations of motion. In fact, its greatest importance lies in two developments. First, the discovery of the symmetrical role of the canonical variables q, p allowed important developments based on the so called *cannonical transformations*, as the Hamilton-Jacobi theory. Second, it also allows a connection, *also of inductive character*, with the microscopic theory of matter, Quantum Mechanics. This connection takes place through the cited canonical formulation of Hamiltonian mechanics, in which the Hamiltonian function acts as the generator of the dynamics. Exploring the symmetry of q and p to its limit, through canonical transformations and Hamilton-Jacobi theory, allowed the induction of a wave equation for matter, through the analogy of classical mechanics with geometric optics. At the same time, the properties of quantities such as Poisson brackets led to the so-called canonical quantization, in which the analogy of these brackets with the quantum commutators is remarkable.

J. R. Mohallem, *Lagrangian and Hamiltonian Mechanics*,
https://doi.org/10.1007/978-3-031-55202-1_3

In this chapter, we follow a practically analogous line to the previous one, focusing mainly on theoretical topics or passages that have proved particularly difficult. to students. On the other hand, to explore the theoretical potential of Hamiltonian mechanics, the emphasis will be shifted from the equations of motion to symmetry properties of transformed canonical pairs and of the Hamiltonian itsel. Also, topics that are most conveniently treated by Lagrangian theory, such as the issue of moving constraints, will be avoided.

A particular emphasis will be given to the topic of *adiabatic invariance.* Real oscillatory systems show deviation from model behavior mainly due to two factors: dissipation and parametric variation. In the last case, the concept of *adiabatic invariant* can become more important than the idealized full invariance in the absence of the perturbation. In the physics of accelerators, for example, this concept is fundamental. For these reasons, we pay special attention here to the theme, which is presented as the natural outlet of the Hamilton-Jacobi theory.

3.1 Canonical Variables and the Hamiltonian Function

Suppose we decide, on a whim for now, to swap the variables $\dot{q}_k$ for their corresponding p_k in our formulation of mechanics. Let us see if we can do it within the Lagrangian formulation. This will be investigated by writing the differential of $L = L(q, \dot{q}, t)$,[1]

$$dL = \sum_{k=1}^{n} [\frac{\partial L}{\partial q_k} dq_k + \frac{\partial L}{\partial \dot{q}_k} d\dot{q}_k] + \frac{\partial L}{\partial t} = \sum_{k=1}^{n} [\dot{p}_k dq_k + p_k d\dot{q}_k] + \frac{\partial L}{\partial t} dt, \qquad (3.1)$$

where we used Lagrange's equation and the definition $p_k = \dfrac{\partial L}{\partial \dot{q}_k}$ for the second step.

But the differential $d\dot{q}_k$ itself is not, in general, proportional to dp_k, except in cases of direct proportionality between p_k and q_k. So, deceptively, our goal of getting free of $\dot{q}$ variables is not achievable.

But we do not give it up! Looking at the definition of h, Eq. (2.98), we see that it resembles a Legendre transformation, conveniently an operation for exchanging dependent and independent variables. Since h has been shown to be an important dynamical quantity, it looks like a good trial to use it trying to accomplish our goal. Our procedure will be to introduce a new function, *the Hamiltonian H* by trying to replace, in $h = h(q, \dot{q}, p, t)$, variables $\dot{q}$ with p. Schematically, the Hamiltonian should be obtained as:

$$h(q, \dot{q}, p, t) \quad \rightarrow \quad H(q, p, t) \qquad (3.2)$$

---

[1] Note that if $Z = Z(x, y)$, then $dZ = \frac{\partial Z}{\partial x} dx + \frac{\partial Z}{\partial y} dy$.

Then, let us take the differential of $h = \sum\limits_{k=1}^{n} p_k \dot{q}_k - L(q, \dot{q}, t)$,

$$dh = \sum_{k=1}^{n}\left[p_k d\dot{q}_k + q_k \dot{p}_k - \frac{\partial L}{\partial q_k}dq_k + \frac{\partial L}{\partial \dot{q}_k}d\dot{q}_k\right] + \frac{\partial L}{\partial t}dt. \tag{3.3}$$

Using again Lagrange's equation and the definition of p_k, we get,

$$dh = \sum_{k=1}^{n}\left[\dot{q}_k dp_k - \dot{p}_k dq_k\right] - \frac{\partial L}{\partial t}dt. \tag{3.4}$$

Voilá, h is really a function of the independent variables q and p (and time, eventually). In fact, under this condition it is h no more, now it is called the Hamiltonian $H(q, p, t)$.

Then, in abstract, being L a function of the independent variables $(q, \dot{q}, t)$, we introduce the *Hamiltonian function* (or, simply, Hamiltonian) $H(q, p, t)$ as being

$$H(q, p, t) = \sum_{k=1}^{n} \dot{q}_k p_k - L(q, \dot{q}, t), \tag{3.5}$$

in which the generalized velocities are replaced by the generalized momenta through the inversion of relation (2.38). An immediate consequence is that, in systems in which the condition $h = E$ is satisfied, it follows immediately that

$$H(q, p, t) = T + V, \tag{3.6}$$

where, in the expression of kinetic energy T, the substitution $\dot{q} \rightarrow p$ above is made.

Now, there is another extraordinary observation from Eq. (3.4): Except for a minus sign, variables q and p are not only independent but also play a symmetric role, in the sense that they could be exchanged in that equation. Nothing like this occurs in Lagrangian theory. That is one of the reasons for they receiving the fancy name *canonical variables*. This symmetry will be more highlighted soon and explored a lot in further developments.

3.2 Hamilton's Equations

The differential of equation (3.5) results as,

$$dH = \sum_{k=1}^{n}(\dot{q}_k dp_k - \dot{p}_k dq_k) - \frac{\partial L}{\partial t}. \tag{3.7}$$

Exercise Prove this result using Lagrange's equation and the definition of p_k.

On the other hand, being $H = H(q, p, t)$, its differential can be also written as

$$dH = \sum_{k=1}^{n} \left(\frac{\partial H}{\partial q_k} dq_k + \frac{\partial H}{\partial p_k} dp_k \right) + \frac{\partial H}{\partial t}. \tag{3.8}$$

Comparing Eqs. (3.7) and (3.8), we get to *Hamilton's equations of motion*,

$$\dot{q}_k = \frac{\partial H}{\partial p_k}, \qquad \dot{p}_k = -\frac{\partial H}{\partial q_k}, \qquad k = 1, n. \tag{3.9}$$

What do these equations mean? Since H depends on variables (q, p) and not on $\dot{q}$, they appear as being $2n$ first-order differential equations of motion for q_k and p_k which must be solved simultaneously, resulting in

$$q_k = q_k(t) \quad e \quad p_k = p_k(t), \tag{3.10}$$

that is, in the solution of the equations of motion.

Also from Eqs. (3.7) and (3.8), we get the relation,

$$\frac{\partial H}{\partial t} = -\frac{\partial L}{\partial t}, \tag{3.11}$$

which expresses the fact that a possible explicit time-dependence of L also applies to H. In this way, $\dfrac{\partial H}{\partial t} \neq 0$ also prevents the invariance of H.

Exercise Show, by direct derivation of $H(q, p, t)$, that

$$\frac{dH}{dt} = \frac{\partial H}{\partial t}, \tag{3.12}$$

which means that, *if H does not depend explicitly on time t, it is then conserved.*

3.2.1   The Phase Space

The space of $2n$ dimensions with one axis for each q_k and one for each p_k is called *phase space*. A point in phase space represents a *state* of the system.[2] The sequence

---

[2] While a point in configuration space gives just the position of each particle, a point in phase space gives positions and momenta.

of points that represents the dynamic evolution of the system in t is a trajectory in phase space, $(q(t), p(t))$, see Fig. 3.1. Of course, analogous to what happens for the configuration space, the trajectory in the phase space can be represented in a parametric form, with one more axis for the parameter t. In this book we use the term "parametric trajectory" for this case. Given the Hamiltonian H, to each set of initial values (q_0, p_0) corresponds a single path in phase space. Different paths can not cross, since this possibility would violate classical determinism.

Example 3.1 Using Lagrangian (1.33), we easily obtain, by definition (Eq. (3.5)),

$$H = e^{-\gamma t}\frac{p^2}{2m} + e^{\gamma t}\frac{k}{2}x^2. \tag{3.13}$$

Using Hamilton's equations (3.9), we get

$$\dot{x} = \frac{\partial H}{\partial p} = \frac{p}{m}e^{-\gamma t}, \tag{3.14}$$

which here corresponds to the expression of the canonical momentum $p = m\dot{x}e^{-\gamma t}$, and

$$\dot{p} = -\frac{\partial H}{\partial x} = -kxe^{\gamma t}. \tag{3.15}$$

These are two coupled equations involving x, p, t with no trivial analytical solution. Substituting one into another, we get $\ddot{x} + \gamma\dot{x} + \dfrac{k}{m}x = 0$, which is the equation of motion (1.36) for the problem. This procedure is nonsensical, however, in the present context. The two Hamilton's equations must be solved separately and simultaneously, in the case by approximate methods, generating a trajectory in the phase space.

Since $\dfrac{\partial H}{\partial t} \neq 0$, H is not conserved. We easily conclude that $H = E$ as well (check the two conditions for that!), so E is not conserved either.

Now, taking the simpler case in which the "potential energy" vanishes in equation (3.13), p has the same expression but now it is a conserved quantity p_0 as x is cyclic. The Hamiltonian becomes $H = e^{-\gamma t}\dfrac{p^2}{2m}$, from which we obtain $\dot{x} = \dfrac{p_0}{m}e^{-\gamma t}$. If the particle starts moving from the origin at $t = 0$, $x = \dfrac{p_0}{m\gamma}(1 - e^{-\gamma t})$. The trajectory in phase space is a horizontal straight line from $x = 0$ to $x_0 = \dfrac{p_0}{m\gamma}$. The physical system is a particle in a resistive medium and the linear momentum $m\dot{x}$ is not conserved, of course.

Exercise Check that, in this case, $\ddot{x} + \gamma\dot{x} = 0$

3.3 What Is Really a Canonical Pair?

Unlike q and $\dot{q}$ in the Lagrangian formulation, the symmetry of the variables q_k and p_k in the Hamiltonian formulation is to be highlighted. The very characterization of q_k as "coordinates" and p_k as "momenta" seem to lose its meaning. This can be recognized by a very simple argument. Being the q_k generalized coordinates and the p_k generalized momenta, let us consider now the change of symbols,

$$q_k \;\; \to \;\; -P_k, \quad p_k \;\; \to \;\; Q_k. \tag{3.16}$$

Taking them to Hamilton's equations, we get

$$\dot{Q}_k = \frac{\partial H}{\partial P_k}, \qquad \dot{P}_k = -\frac{\partial H}{\partial Q_k}. \tag{3.17}$$

Now, looking only at these two equations, it is not possible to state that the Q_k are not generalized coordinates and that the P_k are not generalized moments. In words, the exchange of "coordinates" for "moments" has no consequences on Hamilton's equations. In this formulation, the q_k and the p_k play a symmetrical and equivalent role

Suppose, in a test, a beginning student is asked to obtain the 2nd-order equation of motion for the 1D harmonic oscillator from Hamilton's equations, but make the mistake of switching the coordinate and momentum variables, that is,

$$H = \frac{X^2}{2m} + \frac{k}{2}P^2. \tag{3.18}$$

Hamilton's equations yield

$$\dot{X} = kP \quad e \quad \dot{P} = -\frac{X}{m}. \tag{3.19}$$

Differentiating the first with respect to time and replacing the second, he got $\ddot{X} + \frac{k}{m}X = 0$, the correct equation of the oscillator. A lucky student, no doubt.

Exercise This impressive property is unique to the formulation Hamiltonian. Try do the same thing with Lagrange's equation (you won't make it!).

What we learn here is that the (q, p) variables are "so symmetric" that exchanging their roles in the Hamiltonian does not change the results. So, shouldn't they be called "canonical", or have another fancy name?

3.4 Hamilton's Principle in Phase Space

The reader should have already suspect that Hamilton's equations also follow from a variational principle. In fact, there are many slightly modified versions of Hamilton's principle leading to Hamilton's equations, differing on the form of the integrand and/or on the restrictions on the limiting points. We limit ourselves here to the most common version. Note that now the action depends on two sets of variables, q and p, which must vary independently and unrestrictedly, so the variations of the paths are done in the phase space.

Consider the variation of the action S in the form,

$$\delta S[q, p] = \delta \int_{t_1}^{t_2} \left[\sum_{k=1}^{n} \dot{q}_k p_k - H(q, p, t) \right] dt = 0, \qquad (3.20)$$

where the q_k and the p_k suffer independent variations, but only the q_k remain subject to the restriction $\delta q_k(t_1) = \delta q_k(t_2) = 0$.[3] The form of the integrand is clearly induced by its equivalence to the Lagrangian, but recall that now we are dealing with different variables and space. Performing the variation we obtain,

$$\delta S = \int_{t_1}^{t_2} dt \sum_{k=1}^{n} \left(p_k \delta \dot{q}_k + \dot{q}_k \delta p_k - \frac{\partial H}{\partial q_k} \delta q_k - \frac{\partial H}{\partial p_k} \delta p_k \right) = 0. \qquad (3.21)$$

Integrating the first term by parts,

$$\int_{t_1}^{t_2} p_k \delta \dot{q}_k dt = \int_{t_1}^{t_2} p_k d\delta q_k = [p_k \delta q_k]_{t_1}^{t_2} - \int_{t_1}^{t_2} \delta q_k dp_k = - \int_{t_1}^{t_2} \delta q_k \dot{p}_k dt, \qquad (3.22)$$

so that,

$$\delta S = \int_{t_1}^{t_2} dt \sum_{k=1}^{n} \left[\left(-\dot{p}_k - \frac{\partial H}{\partial q_k} \right) \delta q_k + \left(\dot{q}_k - \frac{\partial H}{\partial p_k} \right) \delta p_k \right] = 0. \qquad (3.23)$$

As the variations δq_k and δp_k are independent, their coefficients must be zero, leading to Hamilton's equations (3.9).

---

[3] In case it went wrong we should try another recipe. In fact, restrictions on the p_k also works, see next exercise.

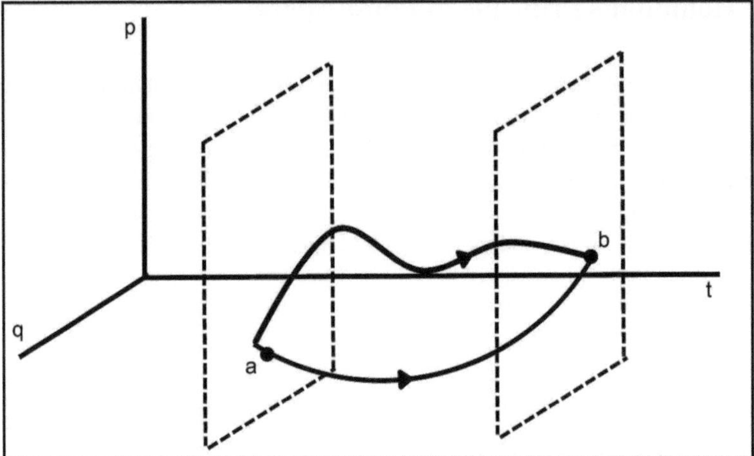

Fig. 3.1 Parametrized trajectories in phase space

Exercise Show that variation of the *generalized action S'* defined as

$$\delta S' = \delta \int_{t_1}^{t_2} \left[\frac{1}{2} \sum_{k=1}^{n} (p_k \dot{q}_k - q_k \dot{p}_k) - H(q, p, t) \right] dt = 0, \qquad (3.24)$$

and subject to $\delta q_k(t_1) = \delta q_k(t_2) = 0$, $\delta p_k(t_1) = \delta p_k(t_2) = 0$ (see Fig. 3.1), also yields Hamilton's equations.

With the integrand and the boundary conditions in the form of the exercise above, the symmetry of coordinates and momenta is fully explicit. But there is another feature that strikes the eye in Eq. (3.24). The integrand of the above equation is no longer, now definitely, a Lagrangian. This equation establishes a variational principle which means no explicit mention of a prior Lagrangian function. On the contrary, it considers generalized coordinates and moments as absolutely independent variables with no need to refer to Eq. (2.38) to define the momenta. This equation can still be the definition of the generalized momenta associated with a generalized coordinate (distance or angle). However, obedience to Hamilton's equations for a pair (q, p) lends it the status of a canonical pair, regardless of Eq. (2.38) and of prior knowledge of the Lagrangian. Later, we will establish a criterion for checking whether a pair of variables is canonical or not, using the Poisson brackets.

We can finally establish Hamilton's principle in phase space as:

Of all possible trajectories in phase space of a system of n degrees of freedom, between instants t_1 and t_2, the one that minimizes the generalized action S, in any of its forms, corresponds to the trajectory actually followed by the system.

3.4.1   Symmetries and Cyclic Coordinates

What are the consequences of the explicit symmetries in L, through cyclic coordinates, on the Hamiltonian theory? Well, we know that if q_l is cyclic, then p_l is constant. Hamilton's equations give

$$\dot{p}_l = -\frac{\partial H}{\partial q_l} = 0 \quad \text{and} \quad \dot{q}_l = \frac{\partial H}{\partial p_l}. \tag{3.25}$$

The second is a first-order differential equation (first integral) in q_l, with constant p_l coming from the first. Out of the canonical pair (q_l, p_l), only q_l will evolve in time. As an illustration, let us return to Example 2.8, already assuming the planar motion. Constructing the Hamiltonian, taking into account that $p_\varphi = l$ is constant, we have,

$$p_r = \frac{\partial L}{\partial \dot{r}} = m\dot{r}, \quad p_\varphi = \frac{\partial L}{\partial \dot{\varphi}} = mr^2\dot{\varphi} = l$$

$$\Rightarrow H = T + V = \frac{p_r^2}{2m} + \frac{l^2}{2mr^2} + V(r). \tag{3.26}$$

This Hamiltonian is written only in terms of r and p_r, while the Lagrangian (2.89) depends on r, $\dot{r}$ and $\dot{\varphi}$. The problem was reduced to two first-order equations,

$$\dot{p}_r = -\frac{\partial H}{\partial r} = -\frac{dV}{dr}, \quad \dot{r} = \frac{\partial H}{\partial r} = \frac{p_r}{m}, \tag{3.27}$$

where the second one just repeats the definition of p_r. Of course φ is not constant, and will be determined by integrating $mr^2\dot{\varphi} = l$, after solving Hamilton's equations for r.

In general, if q_n (chosen as the last coordinate only by convenience) does not appear in the Hamiltonian, its conjugate moment is constant, $p_n = \alpha$, and the Hamiltonian will have the form

$$H = H(q_1, ..., q_{n-1}, p_1, ..., p_{n-1}, \alpha, t), \tag{3.28}$$

which now justifies the alternative term "ignorable" for a cyclic coordinate.

Other relevant consequences of the symmetric form of Hamilton's equations will be explored in the next sections.

3.4.2   Some Examples

Despite being first-order differential equations, Hamilton's equations are not necessarily the best choice for solving usual problems, mainly when analytical solutions are devised. As we have already seen in Example 3.1 and will see next in application to a charge in electromagnetic field, Hamilton's equations can become a little challenging, depending on the form of the generalized momenta. They are more appropriate for more complicated many-body research simulations. However, at least one illustration of an analytic solution of Hamilton's equations is essential to unveil how their mechanism operates, in the next Example. On the other hand, we saw that important information can be extracted from the simple knowledge of the Hamiltonian and its symmetries, so that its determination is always important, regardless of whether we perform the solution of Hamilton's equations or not.

Example 3.2 Here we use the simple 1D harmonic oscillator to explain the "philosophical" difference between applications of Lagrange's and Hamilton's equations. The Lagrangian is simply

$$L = \frac{m}{2}\dot{x}^2 - \frac{k}{2}x^2 \tag{3.29}$$

and generates, by Lagrange's equations, the second order differential equation of motion,

$$\ddot{x} + \omega_0^2 x = 0, \quad \omega_0 = \sqrt{\frac{k}{m}}. \tag{3.30}$$

In turn, the Hamiltonian is obtained as

$$H = xp - L, \quad p = \frac{\partial L}{\partial \dot{x}} = m\dot{x} \quad \Rightarrow \quad H = \frac{p^2}{2m} + \frac{k}{2}x^2 (= E). \tag{3.31}$$

Note, in passing, that for this conservative system, $H = T + V$. Hamilton's equations yield,

$$\dot{x} = \frac{\partial H}{\partial p} = \frac{p}{m} \quad e \quad \dot{p} = -\frac{\partial H}{\partial q} = -kx. \tag{3.32}$$

We then have two first-order equations in the independent variables x and p. They must be resolved simultaneously (usually by numerical methods), starting from initial values x_0 and p_0, thus obtaining the trajectory in phase space. In the present case, however, an analytical solution is possible. Starting from the second equation, writing $\dot{p} = \frac{dp}{dx}\dot{x}$, we get

$$\frac{dp}{dx}\dot{x} = -kx, \quad \int_{p_0}^{p} p\,dp = -mk \int_{x_0}^{x} x\,dx, \tag{3.33}$$

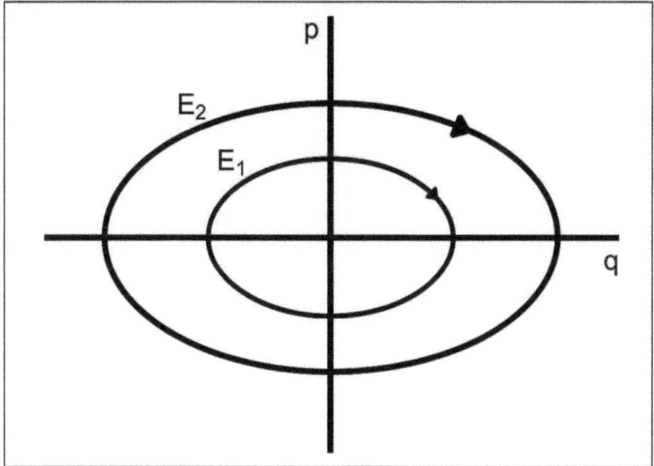

Fig. 3.2 Harmonic oscillations correspond to ellipses in phase space, with $E_2 > E_1$

where we used the first of Hamilton's equations for $\dot{x}$. The separation of variables x and p illustrates the first-order character of the solution. Simple integrations produces,

$$\frac{p^2}{2mE} + \frac{x^2}{(E/2k)} = 1, \tag{3.34}$$

an ellipse in phase space, as expected, see Fig. 3.2. In the figure we show ellipses for different values of the oscillator energy. The student must pay attention to the fact that substituting one Hamilton equation into the other, in order to eliminate p, just restores second-order Lagrange's equation (3.30) and is a meaningless procedure in the present context.

In the following examples, Hamilton's equations are not particularly useful. We will explore these examples especially in the aspects related to the conservation of the Hamiltonian H and its possible identification with the energy of the system.

Example 3.3 The Lagrangian of a non-relativistic charged particle in a electromagnetic field has already been introduced in Eq. (1.51),

$$L = T - U = \frac{m}{2}(\dot{x}^2 + \dot{y}^2 + \dot{z}^2) - q\Phi + \frac{q}{c}\vec{v}\cdot\vec{A}, \tag{3.35}$$

from which we obtain,

$$p_i = m\dot{x}_i + \frac{q}{c}A_{x_i}, \quad i = x, y, z, \quad \Rightarrow \quad \vec{p} = m\vec{v} + \frac{q}{c}\vec{A}. \tag{3.36}$$

Note that the canonical momentum is not equal to the linear momentum $m\vec{v}$. Using this result, the Hamiltonian is found as,

$$H = \vec{v}\cdot\vec{p} - L = \frac{1}{2m}\left(\vec{p} - \frac{q}{c}\vec{A}\right)^2 + q\Phi. \tag{3.37}$$

If the potentials Φ and $\vec{A}$ do not depend explicitly on of t, $\dfrac{\partial H}{\partial t} = -\dfrac{\partial L}{\partial t} = 0$, and we conclude that H is a constant of motion. Independently, realizing that $\vec{p} - \dfrac{q}{c}\vec{A} = m\vec{v}$, we conclude that

$$H = \frac{m}{2}v^2 + q\Phi = E. \tag{3.38}$$

This conclusion may seem contradictory. After all, the potential depends on the particle's velocity. However, realize that the independence of the potential with respect to velocity is a sufficient condition to get $H = E$, but not necessary. In the present case, the magnetic force does no work on the particle (it is perpendicular to the displacement), while the electrostatic force is conservative. Finally, keep the student in mind that this last equation is not useful for obtaining Hamilton's equations of motion, for which the form in Eq. (3.37) must be used.

The first Hamilton's equation takes the form,

$$\dot{\vec{r}} = \frac{\partial H}{\partial \vec{p}} = \frac{1}{m}\left(\vec{p} - \frac{q}{c}\vec{A}\right), \tag{3.39}$$

again simply the redefinition of the generalized moment, while the second, now considering the more general case $\Phi = \Phi(\vec{r},t)$ and $\vec{A} = \vec{A}(\vec{r},t)$, takes the (not necessarily encouraging) form

$$\dot{\vec{p}} = -\frac{\partial H}{\partial \vec{r}} = -\vec{\nabla}H = -\frac{1}{2m}\vec{\nabla}\left(\vec{p} - \frac{q}{c}\vec{A}\right)^2 - q\vec{\nabla}\Phi. \tag{3.40}$$

In the unlikely case of interest in solving such equations, the first term of the right-hand-side of Eq. (3.40) can be manipulated by using the identity $\vec{\nabla}(\vec{B}\cdot\vec{B}) = 2(\vec{B}\cdot\vec{\nabla})\vec{B} + 2\vec{B}\times(\vec{\nabla}\times\vec{B})$.

Equations (3.39)–(3.40) must be solved simultaneously, not substituted one into the other. More instructive is the discussion referring to conservation of H and E, considered also in the following examples.

Example 3.4 We have already obtained the Lagrangian for a particle subject to a scalar field in the relativistic regime, Eq. (1.37). The canonical momentum is obtained simply by

$$\vec{p} = \frac{\partial L}{\partial \vec{v}} = \frac{m\,\vec{v}}{\sqrt{1 - \dfrac{v^2}{c^2}}}, \qquad (3.41)$$

equal to the relativistic linear momentum. So,

$$H = \vec{p} \cdot \vec{v} - L = \frac{mc^2}{\sqrt{1 - \dfrac{v^2}{c^2}}} + V(\vec{r}) = E. \qquad (3.42)$$

Here, with no surprise, $\dfrac{\partial H}{\partial t} = 0$ implies the conservation of H. The use of Cartesian coordinates without constraints, together with the velocity independent potential, guarantee $H = E$, even if $L \neq T - V$. However, the Hamiltonian must be expressed in terms of momenta, not velocities, so that we can use Hamilton's equations. Since the elimination of $\vec{v}$ by inverting Eq. (3.41) is quite complicated, we resort to the expression of the total energy of a relativistic free particle, $\sqrt{p^2 c^2 + m^2 c^4}$, to replace the "kinetic" term in the expression for H, obtaining

$$H = \sqrt{p^2 c^2 + m^2 c^4} + V(\vec{r}). \qquad (3.43)$$

Example 3.5 Finally, for a charged particle in a electromagnetic field, in the non-covariant relativistic regime, the Lagrangian is given by Eq. (1.59), leading to the momentum vector

$$\vec{p} = \frac{\partial L}{\partial \vec{v}} = \frac{m\,\vec{v}}{\sqrt{1 - \dfrac{v^2}{c^2}}} + \frac{q}{c}\vec{A}, \qquad (3.44)$$

which differs from the relativistic linear momentum, Eq. (3.41) by the last term. For the Hamiltonian, we obtain

$$H = \vec{p} \cdot \vec{v} - L = \frac{mc^2}{\sqrt{1 - \dfrac{v^2}{c^2}}} + q\,\Phi. \qquad (3.45)$$

If Φ and $\vec{A}$ are static, H is conserved and, for the same reason explained in Example 3.4, we have $H = E$. Finally, the Hamiltonian in terms of canonical momenta is

$$H = \sqrt{(\vec{p} - \frac{q}{c}\vec{A})^2 c^2 + m^2 c^4} + q\Phi. \tag{3.46}$$

Again, solving Hamilton's equations in both last examples is not an easy task, but, as already discussed, it is not our main interest. In addition to checking for conservation of E and H, which in the relativistic cases assume forms that differs from the traditional sum of kinetic and potential energies, the expressions of the Hamiltonian itself in terms of the canonical variables have fundamental importance in the quantum theory of matter, where H plays a central role

Exercise The length r of a simple pendulum varies at the constant rate $r = l - \alpha t$, where l is a constant. Find H, derive Hamilton's equations and discuss the conservation of H and the E.

3.5 Canonical Transformations

Smart transformations of dynamical variables can simplify the solution of problems in mechanics. Within Lagrangian theory, we have already seen that choosing generalized coordinates containing one or more cyclic coordinates makes the solution easier. This procedure is always allowed since Lagrange's equations are invariant under point transformations in the configuration space, see Sect. 2.3.1. This means that the chosen transformation equations and the relation $\dot{q}_k = \dfrac{dq_k}{dt}$ guarantee the invariance of Lagrange's equations. In turn, there is no equivalent feature in Hamilton's mechanics. The canonical variables (q, p) are fully independent of each other, so that acceptable transformations must obey some criteria. On the other hand, transformations known as *canonical*, defined in the sequence, are a powerful and general tool in Hamiltonian mechanics.

3.5.1 General Transformations

A generalized coordinate change,

$$Q_k = Q_k(q_1, ..., q_n, t), \tag{3.47}$$

is known as *point transformation* in the configuration space. Let us go back to the example of the particle under the action of a central force, see Example 2.7. The Lagrangian in Cartesian coordinates,

$$L = \frac{m}{2}(\dot{x}^2 + \dot{y}^2 + \dot{z}^2) - V(\sqrt{x^2 + y^2 + z^2}), \tag{3.48}$$

does not exploit the spherical symmetry of the potential. In spherical coordinates, on the other hand, the Lagrangian in the form of Eq. (2.86) makes this symmetry explicit and, consequently, yields the reduction of the number of generalized coordinates to two, even in the absence of constraints. Moreover, in the reduced Lagrangian, Eq. (2.89), φ is still cyclical, evidencing the existence of a first integral, p_φ, Eq. (2.87). The equation in φ was reduced to first-order, while the equation in r is still of second order. This is as much of simplification as you can get in the Lagrange's formulation, due to the fact that Lagrange's equations are of second order. Independently, we have shown in Sect. 2.3.1 that the *form* of Lagrange's equations is maintained whatever the set of coordinates used, that is,

$$\frac{d}{dt}\frac{\partial L}{\partial \dot{q}_k} - \frac{\partial L}{\partial q_k} = 0 \quad \Rightarrow \quad \frac{d}{dt}\frac{\partial L}{\partial \dot{Q}_k} - \frac{\partial L}{\partial Q_k} = 0, \quad k = 1, ..., n. \tag{3.49}$$

A consequence is that the Lagrangian $L(Q, \dot{Q}, t)$ can be obtained from $L(q, \dot{q}, t)$ by direct replacement of the transformation variables (3.47) in it. At least in the absence of constraints, the very procedure of writing L in Cartesian coordinates and, in the next step, transforming it in function of the chosen generalized coordinates, already carries this assumption. In conservative systems, such that we can write $L = T - V$ and $H = T + V$, the same procedure for the Hamiltonian is obviously valid, but generates $H(Q, \dot{Q}, t)$ and not $H(Q, P, t)$, as required.

As we saw that in the Hamiltonian treatment there is an equivalence between coordinates and momenta, it seems natural to extend the idea of point transformation to the $2n$ variables of the phase space. A *point transformation in phase space* consists of exchanging q_k and p_k for other sets Q_k and P_k, such that

$$q_k \quad \rightarrow \quad Q_k = Q_k(q, p, t), \qquad p_k \quad \rightarrow \quad P_k = P_k(q, p, t). \tag{3.50}$$

However, direct substitution of the new variables in H does not work in general. This is because, given the symmetry between coordinates q_k and momenta p_k, we want the new variables to depend on both sets of old variables, as stated in Eq. (3.50) above. Explaining better, referring to the point transformation (3.47), the corresponding momenta are,

$$P_k = \frac{\partial L}{\partial \dot{Q}_k} = \sum_{l=1}^{n} \frac{\partial L}{\partial \dot{q}_l}\frac{\partial \dot{q}_l}{\partial \dot{Q}_k} = \sum_{l=1}^{n} a_{lk} p_l, \tag{3.51}$$

that is, they are linear combinations of the original momenta p_k. In transformation like equation (3.50), no such or any other previous dependence of P_k on p_k are assumed. What is demanded for the transformation is that they obey Hamilton's equations in the phase space (Q, P). This being the case, we refer to them as *canonical transformations*, which thus are *point transformations in phase space that preserve the form of Hamilton's equations relative to the transformed Hamiltonian.*

This demand can be kept in two ways which, although not independent (the second contains the first), should be presented separately for optimal understanding by beginners. For simplicity, from now on we will refer to a point transformation only concerning the configuration space. Transformations in phase space of the type (3.50) will always be referred to as canonical transformations.

3.5.2 Canonical Transformation with $K = H$ (Direct Substitution)

Assuming that Eqs. (3.50) are invertible, so that we can have $q_k = q_k(Q, P, t)$ and $p_k = p_k(Q, P, t)$, we introduce the new Hamiltonian, the *Kamiltonian*[4] $K(Q, P, t)$ as

$$K(Q, P, t) = H(q(Q, P, t), p(Q, P, t), t), \tag{3.52}$$

that is, by direct substitution of the new variables in H. However, we need to ensure that our transformation is canonical, applying Hamilton's principle in the new phase space (Q, P), which would then be stated as

$$\delta \int_{t_1}^{t_2} \left[\sum_{k=1}^{n} \dot{Q}_k P_k - K(Q, P, t) \right] dt = 0. \tag{3.53}$$

Since we are working with $K = H$, the sufficient condition for the simultaneous validity of the variations (3.20) and (3.53) is that the sums $\sum_{k=1}^{n} \dot{q}_k p_k$ and $\sum_{k=1}^{n} \dot{Q}_k P_k$ differ at most by the total time derivative of an arbitrary function $F(q, p)$, or equivalently,

$$\sum_{k=1}^{n} (p_k dq_k - P_k dQ_k) = dF(q, p), \tag{3.54}$$

In fact, under this condition, according to Hamilton's principle,

$$\delta \int_{t_1}^{t_2} \left[\sum_{k=1}^{n} \dot{q}_k p_k - H(q, p, t) \right] dt = \delta \int_{t_1}^{t_2} \left[\sum_{k=1}^{n} \dot{Q}_k P_k - K(Q, P, t) \right] dt$$

$$+ \delta \int_{t_1}^{t_2} dF = 0. \tag{3.55}$$

---

[4] This funny name seems to have been used for the first time in Goldstein's book, see bibliography.

As the last term is evidently zero, Hamilton's principle applied to the phase space (Q, P) preserves the form of Hamilton's equations in that space. Equation (3.53) then holds, so that,

$$\dot{Q}_k = \frac{\partial K}{\partial P_k}, \qquad \dot{P}_k = -\frac{\partial K}{\partial Q_k}, \tag{3.56}$$

which are Hamilton's equations in the transformed phase space.

A helpful observation is that, as the integrand of Eq. (3.55) differ only by terms of the type $\sum \dot{q}p$, then F does not depend explicitly on time, $\frac{\partial F}{\partial t} = 0$, for $K = H$.

Note also that there is an apparent ambiguity in Eq. (3.54), since its left side involves differentials of q_k and Q_k, while its right side is a differential of a function of q_k and p_k. This is not a problem because we can, in principle, use the canonical transformations to interchange variables, say $p = p(q, Q, t)$, in $F(q, p)$. In fact, one can perceive, from the same equation, that information about half of the canonical variables (any pair) might be obtained from the partial derivatives of F, relative to the other half (q, Q). These points will be discussed in more detail in the next section. For now, we can agree in keeping F undefined in what concerns its arguments, or even keeping $F = F(q, p)$.

Example 3.6 The simplest problem of a free particle illustrates well the previous ideas. The Hamiltonian is

$$H = \frac{p^2}{2m}, \tag{3.57}$$

and we propose the canonical transformation (later we will learn to check in advance if a transformation is canonical or not)

$$P = \frac{p^2}{2m}, \quad \text{and} \quad Q = m\frac{q}{p}. \tag{3.58}$$

Inverting these relations, we construct the Kamiltonian K,

$$p = \sqrt{2mP}, \quad q = \frac{Q}{m}\sqrt{2mP}, \quad K(= H) = P. \tag{3.59}$$

The transformed Hamiltonian is equal to the transformed momentum P! Once executed the canonical transformation, we must abandon any idea of a standard form for K involving its variables Q and P. The only requirement is that Hamilton's equations hold, which is explicit in the very concept of canonical transformation.

Hamilton's equations yield,

$$\dot{P} = -\frac{\partial K}{\partial Q} = 0 \;\Rightarrow\; P = P_0; \quad \dot{Q} = \frac{\partial K}{\partial P} = 1 \;\Rightarrow\; Q = t + Q_0. \tag{3.60}$$

Taking this result in the expression for q in Eq. (3.58),

$$q = \frac{1}{m}\sqrt{2m\,P_0}(t + Q_0) = \frac{p_0}{m}t + q_0, \quad p = p_0, \tag{3.61}$$

which is the expected solution. In the present example, where the canonical transformation was obtained, say, by inspection, knowledge of the function F was not necessary. However, it is instructive to obtain it as a function of q and p, as suggested above. Using Eqs. (3.54) and (3.58),

$$p\dot{q} - P\dot{Q} = p(\frac{p\dot{Q} + Q\dot{p}}{m}) - \frac{p^2\dot{Q}}{2m} = \frac{p^2\dot{Q}}{2m} + \frac{pQ\dot{p}}{m} = \frac{d}{dt}(\frac{p^2 Q}{2m}), \tag{3.62}$$

that is,

$$F = \frac{p^2 Q}{2m} = \frac{pq}{2} = m\frac{q^2}{Q}, \tag{3.63}$$

the last passage showing that F can really be written in terms of (q, Q), as suggested in the left side of Eq. (3.54).

3.5.3   Canonical Transformation with $K \neq H$

One of the motivations (but not the only one, as we shall see) for performing a canonical transformation has a practical goal: To make $K(Q, P, t)$ simpler than $H(q, p, t)$, so that Hamilton's equations become easier to solve. Evidently, the biggest possible simplification would be to find a canonical transformation leading to

$$K(Q, P, t) = 0, \quad \text{or} \quad K(Q, P, t) = \text{constant}. \tag{3.64}$$

The student should not, at this point, be surprised that a Hamiltonian can be zero or constant. The energy of a conservative system, $E = T + V$ can be also zero, even though T and V vary in time. Note that, in this case, the relation between K and H is no longer previously known; the canonical transformation that generates the expected result for K has to be guessed for now (in the next section we will present a possible solution to this problem).

Being $H \neq K$, the requirement of invariance of the form of Hamilton's equations now proceed by demanding that the full integrands of the action variation in both phase spaces, differ at most by a function $F(q, p, t)$. That is, instead of (3.54), we have

$$\sum_{k=1}^{n} \dot{q}_k p_k - H(q, p, t) - \left[\sum_{k=1}^{n} \dot{Q}_k P_k - K(Q, P, t)\right] = \frac{dF(q, p, t)}{dt}. \tag{3.65}$$

This condition is equivalent of demanding the variations of the action in the two phase spaces to be equal. Note that now F can be explicitly dependent of t.

Rearranging the above equation,

$$\sum_{k=1}^{n}(p_k dq_k - P_k dQ_k) + (K - H)dt = dF. \tag{3.66}$$

F is called *generating function* of the canonical transformation, since the canonical variables not contained in F will be obtained through partial derivatives of it. The above equation characterizes, in fact, a transformation $(q, p) \to (Q, P)$ as being canonical.

We can notice from the term on its left that, in order to be meaningful, it must contain F written in terms of the pair (q, Q). On the other hand, in the transformation (3.50) we can solve the p_k in terms of (q, Q, t) so that F becomes a function of these $2n+1$ variables. Now it will obey Eq. (3.66) without any ambiguity and will be called $F_1(q, Q, t)$. From Eq. (3.66), the other variables will be obtained as,

$$p_k = \frac{\partial F_1}{\partial q_k}, \quad P_k = -\frac{\partial F_1}{\partial Q_k}, \quad k = 1, \ldots, n, \tag{3.67}$$

and the transformed Hamiltonian K becomes,

$$K(Q, P, t) = H(q, p, t) + \frac{\partial F_1}{\partial t}. \tag{3.68}$$

Example 3.7 We resort again to the free particle, with Hamiltonian given by Eq. (3.57), but now choosing the generating function $F_1(q, Q, t) = \dfrac{m(q - Q)^2}{2t}$. Using Eqs. (3.67) above,

$$p = \frac{\partial F_1}{\partial q} = \frac{m(q - Q)}{t}, \quad P = -\frac{\partial F_1}{\partial Q} = \frac{m(q - Q)}{t} = p, \tag{3.69}$$

or, making explicit the transformed variables in terms of the old ones,

$$Q = q - \frac{p}{m}t, \quad P = p. \tag{3.70}$$

In turn, for the transformed Hamiltonian, we easily obtain

$$K = H + \frac{\partial F_1}{\partial t} = 0. \tag{3.71}$$

One more time, it should not be a surprise that a Hamiltonian has zero value. By now, you should be aware that $K(Q, P, t)$ has not any connection with a Lagrangian written in terms of $(Q, \dot{Q}, t)$. The equations of motion result as,

$$\dot{Q} = \frac{\partial K}{\partial P} = 0 \;\Rightarrow\; Q = Q_0, \quad \dot{P} = -\frac{\partial K}{\partial Q} = 0 \;\Rightarrow\; P = P_0, \qquad (3.72)$$

which leads, using Eqs. (3.69) and (3.70), to the solution (3.61) of the problem in the original variables.[5]

It is easy to see that $F_1(q, Q, t)$ cannot generate all desired canonical transformations. Derivatives of F_1 in Eq. (3.67) can only generate functions of q and Q. Just as an example, the identity transformation

$$Q_k = q_k, \quad P_k = p_k \qquad (3.75)$$

cannot have F_1 as its generating function, since the second identity above, $P_k = p_k$, cannot be obtained from a F_1-type function. On the other hand, we should be able to get at least three more fundamental generating functions (for one-particle problems), yielding Table 3.1 bellow.

We see that the identity transformation (3.75) is, instead, generated by $F_2 = \sum_{k=1}^{n} q_k P_k$, while $F_1 = \sum_{k=1}^{n} q_k Q_k$ generates the exchange of coordinates and moments,

Table 3.1 The four fundamental generator functions (one-particle)

$F = F_1(q, Q, t)$	$p_k = \dfrac{\partial F_1}{\partial q_k}, \quad P_k = -\dfrac{\partial F_1}{\partial Q_k}$
$F = F_2(q, P, t) - \sum_k Q_k P_k$	$p_k = \dfrac{\partial F_2}{\partial q_k}, \quad Q_k = \dfrac{\partial F_2}{\partial P_k}$
$F = F_3(p, Q, t) + \sum_k q_k p_k$	$q_k = -\dfrac{\partial F_3}{\partial p_k}, \quad P_k = -\dfrac{\partial F_3}{\partial Q_k}$
$F = F_4(p, P, t) + \sum_k q_k p_k - \sum_k Q_k P_k$	$q_k = -\dfrac{\partial F_4}{\partial p_k}, \quad Q_k = \dfrac{\partial F_4}{\partial P_k}$

---

[5] The student must be careful not to understand Eqs. (3.69) and (3.70) above as signs of the existence of partial time derivatives of the canonical variables. What happens, say, in Eq. (3.70), is that it wouldn't be correct to differentiate Q with respect to t keeping q and p fixed, because the original variables and the transformed ones are interdependent, through the canonical transformation, although they are independent of each other, as discussed in Sect. 1.1.1. There should be no doubt about the meaninglessness of writing

$$\frac{\partial q}{\partial t}, \quad \frac{\partial p}{\partial t}, \quad \frac{\partial Q}{\partial t}, \quad \frac{\partial P}{\partial t}, \qquad (3.73)$$

which, according to the definitions (1.2) and (1.3), are fully equivalent to the total derivatives,

$$\frac{dq}{dt}, \quad \frac{dp}{dt}, \quad \frac{dQ}{dt}, \quad \frac{dP}{dt}. \qquad (3.74)$$

$$Q_k = p_k, \quad P_k = -q_k, \tag{3.76}$$

discussed in Sect. 3.3.

Note that only $F_1(q, Q)$ can be obtained by direct substitution $F_1(q, Q) \leftarrow F(q, p(q, Q, t)t)$. For the other we need to change the first term at left in Eq. (3.54) or (3.66). This will be done using the definitions in the first column of the table. As an illustration, let us obtain the derivative relations involving F_2. Making explicit the variables of each term,

$$F(q, p, t) = F_2(q, P, t) - \sum_{k=1}^{n} Q_k P_k, \tag{3.77}$$

from which

$$dF = \sum_{k=1}^{n} (\frac{\partial F_2}{\partial q_k} dq_k + \frac{\partial F_2}{\partial P_k} dP_k - P_k dQ_k - Q_k dP_k) + \frac{\partial F_2}{\partial t} dt, \text{ or,}$$

$$dF = \sum_{k=1}^{n} [\frac{\partial F_2}{\partial q_k} dq_k - P_k dQ_k + (\frac{\partial F_2}{\partial P_k} - Q_k) dP_k] + \frac{\partial F_2}{\partial t} dt. \tag{3.78}$$

Comparing with Eq. (3.66), we obtain the derivatives $p_k = \dfrac{\partial F_2}{\partial q_k}$, $Q_k = \dfrac{\partial F_2}{\partial Q_k}$ and the relationship between K and H, $K(Q, P, t) = H(q, p, t) + \dfrac{\partial F_2}{\partial t}$. Clearly, the last relation can be generalized as

$$K = H + \frac{\partial F_i}{\partial t}, \quad i = 1, 4. \tag{3.79}$$

Exercise Repeat the procedure above to obtain de derivatives of F_3 and F_4 in Table 3.1.

Example 3.8 Let us go back to Example 3.7 where we made $K = P$ for the free particle, but now in an inverted way, that is, starting from the generating function (3.63), which suggests for F_3,

$$F_3(q, P, t) = F - qp = -\frac{pq}{2} = -\frac{p^2 Q}{2m}, \tag{3.80}$$

We have, from Table 3.1,

$$q = -\frac{\partial F_3}{\partial p} = \frac{pQ}{m}, \quad P = -\frac{\partial F_3}{\partial Q} = \frac{p^2}{2m}. \tag{3.81}$$

In words, knowledge of the generating function leads to the canonical transformation. Moreover,

$$K = H + \frac{\partial F_3}{\partial t} = H.$$ (3.82)

The rest of the solution follows the same steps of Example 3.7, starting from the solution of Hamilton's equations for K. Of course, the method of canonical transformations is to be applied to problems of real difficulty.

Exercise Prove that the relation that defines two equivalent Lagrangians, Eq. (1.62), corresponds to a canonical transformation and find the generating function.

3.5.4 Poisson's Brackets

Being $F(q, p, t)$ a generic function of dynamic variables and time,[6] its time derivative is

$$\dot{F} = \sum_{k=1}^{n} \left(\frac{\partial F}{\partial q_k} \dot{q}_k + \frac{\partial F}{\partial p_k} \dot{p}_k \right) + \frac{\partial F}{\partial t}.$$ (3.83)

Using Hamilton's equation, we get

$$\dot{F} = \sum_{k=1}^{n} \left(\frac{\partial F}{\partial q_k} \frac{\partial H}{\partial p_k} - \frac{\partial F}{\partial p_k} \frac{\partial H}{\partial q_k} \right) + \frac{\partial F}{\partial t}.$$ (3.84)

We will see that the quantity in parentheses plays an important role in the theory. Let us define the *Poisson brackets* of the variables F and G as

$$[F, G]_{q,p} = \sum_{k=1}^{n} \left(\frac{\partial F}{\partial q_k} \frac{\partial G}{\partial p_k} - \frac{\partial F}{\partial p_k} \frac{\partial G}{\partial q_k} \right),$$ (3.85)

so that Eq. (3.84) becomes

$$\dot{F} = [F, H]_{q,p} + \frac{\partial F}{\partial t}.$$ (3.86)

A consequence is that, *if a dynamical quantity does not depend explicitly on time, its variation rate is equal to its Poisson bracket with the Hamiltonian.*

It is easily verified that the fundamental Poisson brackets obey the relations

---

[6] Not to be confused with the generating function of canonical transformations.

$$[q_j, q_l]_{q,p} = 0, \quad [p_j, p_l]_{q,p} = 0, \quad [q_j, p_l]_{q,p} = \delta_{jl}, \quad [p_j, q_l]_{q,p} = -\delta_{jl}.$$

$$(3.87)$$

Let us prove the last relation. From the definition, Eq. (3.85),

$$[q_j, p_l]_{q,p} = \sum_{k=1}^{n} \left(\frac{\partial q_j}{\partial q_k} \frac{\partial p_l}{\partial p_k} - \frac{\partial q_j}{\partial p_k} \frac{\partial p_l}{\partial q_k} \right) = \sum_{k=1}^{n} \frac{\partial q_j}{\partial q_k} \frac{\partial p_l}{\partial p_k} = \delta_{jl}, \qquad (3.88)$$

where we used the fact that the canonical variables are independent from each other, so $\frac{\partial q_k}{\partial p_k} = 0$, and vice-versa, while the remaining crossing derivatives only survive for $k = l = j$.

A fundamentally property of Poisson brackets is their invariance under canonical transformations. Instead of resorting to the *symplectic* structure of Hamiltonian mechanics[7] to perform the proof, we adopt here an alternative approach with help from the fundamental generating functions. Let us prove first the invariance of the fundamental Poisson brackets. Starting from the generic canonical transformation (3.50), we have

$$[Q_j, P_l]_{q,p} = \sum_{k=1}^{n} \left(\frac{\partial Q_j}{\partial q_k} \frac{\partial P_l}{\partial p_k} - \frac{\partial Q_j}{\partial p_k} \frac{\partial P_l}{\partial q_k} \right). \qquad (3.89)$$

Using the generator function $F_3(p, Q)$, we get

$$q_k = -\frac{\partial F_3}{\partial p_k} \quad \Rightarrow \quad \frac{\partial q_k}{\partial Q_l} = -\frac{\partial^2 F_3}{\partial Q_l \partial p_k} = \frac{\partial}{\partial p_k}\left(-\frac{\partial F_3}{\partial Q_l}\right) = \frac{\partial P_l}{\partial p_k}. \qquad (3.90)$$

Now, using $F_1(q, Q)$,

$$p_k = \frac{\partial F_1}{\partial q_k} \quad \Rightarrow \quad \frac{\partial p_k}{\partial Q_l} = \frac{\partial^2 F_1}{\partial Q_l \partial q_k} = \frac{\partial}{\partial q_k}\left(\frac{\partial F_1}{\partial Q_l}\right) = -\frac{\partial P_l}{\partial q_k}. \qquad (3.91)$$

Taking the expressions of $\frac{\partial P_l}{\partial p_k}$ and $\frac{\partial P_l}{\partial q_k}$ into the Poisson's brackets (3.89), we get

$$[Q_j, P_l]_{q,p} = \sum_{k}^{n} \left(\frac{\partial Q_j}{\partial q_k} \frac{\partial q_k}{\partial Q_l} + \frac{\partial Q_j}{\partial p_k} \frac{\partial p_k}{\partial Q_l} \right) = \frac{\partial Q_j}{\partial Q_l}$$

$$= \delta_{jl} = [Q_j, P_l]_{Q,P} = -[P_j, Q_l]_{Q,P}. \qquad (3.92)$$

Similarly, using relations analogous to (3.90)–(3.91), deduced from the F_i , we obtain

---

[7] Otherwise very important for more advanced studies.

$$[Q_j, Q_l]_{q,p} = [Q_j, Q_l]_{Q,P} = 0 \quad \text{and} \quad [P_j, P_l]_{q,p} = [P_j, P_l]_{Q,P} = 0, \qquad (3.93)$$

which then proves the invariance of the fundamental brackets under a canonical transformation. Conversely, the variables Q_k and P_k that obey relations (3.93) are connected with the original variables q_k and p_k by the derivative relations of the second column of Table 3.1. Therefore, we can conclude that: *Obeying relations (3.92) and (3.93) is a necessary and sufficient condition for a transformation to be canonical.*

These relationships can then be used as a test of "canonicity" of a proposed transformation. As a consequence of the exposed above, the canonicity of a transformation like (3.50) is *independent of the particular physical problem* in study. The form of the Hamiltonian may suggest the canonical transformation to be used, but it will be applicable to another problem with the same number of degrees of freedom, whose Hamiltonian displays also the same symmetries.

Example 3.9 Given the model Hamiltonian,

$$H = \frac{p_1^2}{2m} + \frac{(p_2 - kq_1)^2}{2m}, \qquad (3.94)$$

in order to simplify it, the following transformation is suggested,

$$Q_1 = k^{-1}p_1, \quad P_1 = p_2 - kq_1, \quad Q_2 = q_2 - k^{-1}p_1 \text{ e } P_2 = p_2. \qquad (3.95)$$

Let us first check its canonicity. We easily get

$$[Q_1, P_1] = k^{-1}k = 1, \ [Q_1, P_2] = 0, \ ..., \ [Q_2, Q_2] = 0 \ \text{ etc.,} \qquad (3.96)$$

confirming the canonical character of the proposed transformation. From the expressions for the derivatives of any generating function in Table 3.1, we verify that the absence of explicit dependence on time of the transformed variables, implies that the generating function is also not explicitly dependent on time. The transformed Hamiltonian is then be $K(Q, P, t) = H(q(Q, P), p(Q, P), t)$. Substituting into H the original variables for those transformed, we find

$$K = \frac{1}{2m}P_1^2 + \frac{k^2}{2m}Q_1^2, \qquad (3.97)$$

which we identify as the Kamiltonian of a harmonic oscillator. Hamilton's equations yield,

$$\dot{Q}_1 = \frac{\partial K}{\partial P_1} = \frac{P_1}{2m}, \ \ddot{Q}_1 = \frac{\dot{P}_1}{2m} = -\frac{1}{2m}\frac{\partial K}{\partial Q_1}, \text{ or, } \ddot{Q}_1 + \Omega^2 Q_1 = 0, \qquad (3.98)$$

with $\Omega = \dfrac{k}{m}$. The known solutions are

$$Q_1 = A\cos(\Omega t + \delta) \quad \Rightarrow \quad P_1 = m\dot{Q}_1 = -m\Omega A sen(\Omega t + \delta). \tag{3.99}$$

Also,

$$\dot{P}_2 = -\frac{\partial K}{\partial Q_2} = 0 \text{ e } \dot{Q}_2 = -\frac{\partial K}{\partial P_2} = 0 \quad \Rightarrow \quad P_2 = P_{2,0} \text{ e } Q_2 = Q_{2,0}. \tag{3.100}$$

Now, we can get the solutions for the original variables as,

$$q_1 = \frac{P_2 - P_1}{k} = \frac{P_{2,0}}{k} + A sen(\Omega t + \delta), \quad p_1 = kQ_1 = kA\cos(\Omega t + \delta),$$

$$q_2 = Q_2 + \frac{p_1}{k} = Q_{2,0} + A\cos(\Omega t + \delta), \quad p_2 = P_{2,0}. \tag{3.101}$$

The solutions are given in terms of four constants, A, δ, $Q_{2,0}$ and $P_{2,0}$, as expected, since we have four independent variables.

We continue with the proof of the invariance of the Poisson brackets of two any variables. Exploring the inversion of the canonical transformation (3.50)

$$q_k = q_k(Q, P, t) \quad e \quad p_k = p_k(Q, P, t), \tag{3.102}$$

we have

$$[F, G]_{q,p} = \sum_{l=1}^{n} \left(\frac{\partial F}{\partial q_l} \frac{\partial G}{\partial p_l} - \frac{\partial F}{\partial p_l} \frac{\partial G}{\partial q_l} \right) \tag{3.103}$$

$$= \sum_{l,k=1}^{n} \left[\frac{\partial F}{\partial q_l} \left(\frac{\partial G}{\partial Q_k} \frac{\partial Q_k}{\partial p_l} + \frac{\partial G}{\partial P_k} \frac{\partial P_k}{\partial p_l} \right) - \frac{\partial F}{\partial p_l} \left(\frac{\partial G}{\partial Q_k} \frac{\partial Q_k}{\partial q_l} + \frac{\partial G}{\partial P_k} \frac{\partial P_k}{\partial q_l} \right) \right]$$

$$= \sum_{k=1}^{n} \left[\frac{\partial G}{\partial Q_k} \sum_{l=1}^{n} \left(\frac{\partial F}{\partial q_l} \frac{\partial Q_k}{\partial p_l} - \frac{\partial F}{\partial p_l} \frac{\partial Q_k}{\partial q_l} \right) + \frac{\partial G}{\partial P_k} \sum_{l=1}^{n} \left(\frac{\partial F}{\partial q_l} \frac{\partial P_k}{\partial p_l} - \frac{\partial F}{\partial p_l} \frac{\partial P_k}{\partial q_l} \right) \right],$$

or,

$$[F, G]_{q,p} = \sum_{k=1}^{n} \left(\frac{\partial G}{\partial Q_k} [F, Q_k]_{q,p} + \frac{\partial G}{\partial P_k} [F, P_k]_{q,p} \right). \tag{3.104}$$

Making $F = Q_l$,

$$[Q_l, G]_{q,p} = \sum_{k=1}^{n} \left(\frac{\partial G}{\partial Q_k} [Q_l, Q_k]_{q,p} + \frac{\partial G}{\partial P_k} [Q_l, P_k]_{q,p} \right) = \frac{\partial G}{\partial P_k} \delta_{lk} = \frac{\partial G}{\partial P_l},$$

$$\tag{3.105}$$

in view of relations (3.93), so that, by a similar procedure for $[P_l, G]_{q,p}$, we obtain

$$[P_l, G]_{q,p} = -\frac{\partial G}{\partial Q_l}. \tag{3.106}$$

Finally, returning to Eq. (3.104) above, we finish the proof,

$$[F, G]_{q,p} = \sum_{k=1}^{n} \left(-\frac{\partial G}{\partial Q_k} \frac{\partial F}{\partial P_k} + \frac{\partial G}{\partial P_k} \frac{\partial F}{\partial Q_k} \right) = [F, G]_{Q,P}. \tag{3.107}$$

Using the proved invariance, we can take the Poisson brackets of F and G relative to the transformed variables to check relations (3.93), demonstrating the consistency of the results obtained. Moreover, the equality (3.107) indicates that the use of subscripts is unnecessary, so that they will be dropped from now on.

Exercise Check the conditions under which q and $A\dot{q}$ form a canonical pair (perhaps you did know the answer, already).

3.6 Infinitesimal Canonical Transformations and Temporal Evolution

So far we have considered canonical transformations by which a point in phase space (q, p) is mapped to a point in another phase space (Q, P), see Fig. 3.3a.

Any dynamic property $A(q, p, t_f)$ can be written as the transformed property $A(Q, P, t_f)$, at any fixed time t_f. Such a canonical transformation is called *passive*. In turn, an *active* canonical transformation connects points of the same phase space, that is, it maps a point (q, p) to another point (Q, P), both contained in the same phase space, linked by the variation of a continuous parameter, Fig. 3.3b. Surely, it must obey Hamilton's equation in order to be canonical.

Let us consider an active canonical transformation, in which this parameter is the time t. Our goal is to connect points on the trajectory of a dynamical system as t evolves. We have seen that the generating function $F_2 = \sum_k q_k P_k$ yields the identity transformation $Q_k = q_k$ and $P_k = p_k$. A generating function that produces infinitesimal variations in the original canonical variables,

$$Q_k = q_k + \delta q_k \quad \text{e} \quad P_k = p_k + \delta p_k, \tag{3.108}$$

might have a form like

$$F_2 = \sum_{l=1}^{n} q_l P_l + \epsilon G(q, P), \tag{3.109}$$

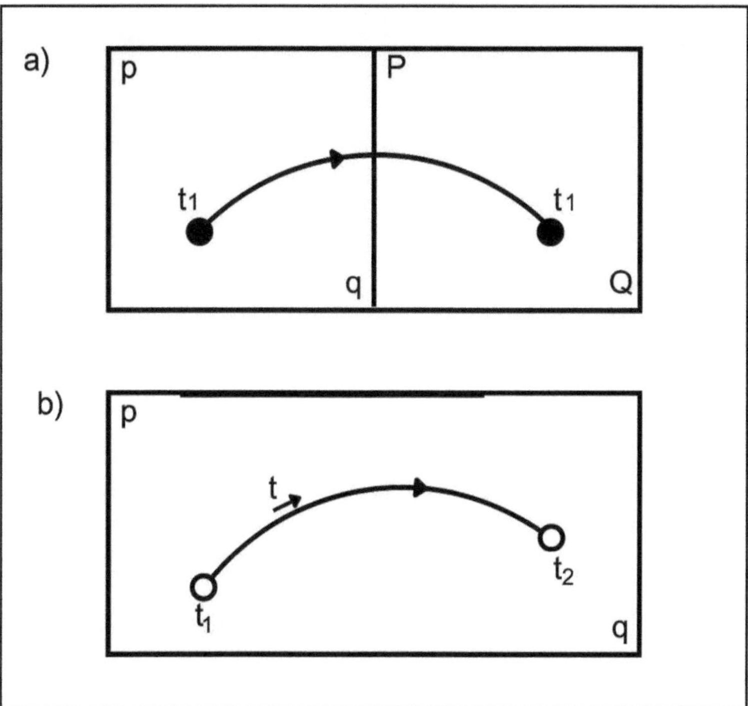

Fig. 3.3 (**a**) Passive and (**b**) active canonical transformations

where ϵ is an infinitesimal parameter. Note that the symbol δA here means the change in quantity A when subjected to the considered canonical transformation (not to be confused with the functional variation). From Table 3.1, we get

$$p_k = \frac{\partial F_2}{\partial q_k} = P_k + \epsilon \frac{\partial G}{\partial q_k}, \quad Q_k = \frac{\partial F_2}{\partial P_k} = q_k + \epsilon \frac{\partial G}{\partial P_k}. \tag{3.110}$$

To consider an active transformation, however, we must replace P with p in the function $G(q, P)$, using the canonical transformation, so to have $G = G(q, p)$. In the second expression in Eq. (3.110), the derivative relative to P_k becomes then inconvenient, but the term $\epsilon \dfrac{\partial G}{\partial P_k}$ in Eq. (3.110) can be transformed as,

$$\epsilon \frac{\partial G}{\partial P_k} = \epsilon \left(\frac{\partial P_k}{\partial G}\right)^{-1} = \epsilon \left[\frac{\partial (p_k + \delta p_k)}{\partial G}\right]^{-1} = \epsilon \frac{\partial G}{\partial p_k}\left[1 + \frac{\partial \delta p_k}{\partial p_k}\right]^{-1}$$

$$\simeq \epsilon \frac{\partial G}{\partial p_k}\left[1 - \frac{\partial \delta p_k}{\partial p_k}\right] \simeq \epsilon \frac{\partial G}{\partial p_k}. \tag{3.111}$$

In the last passage, the derivative $\dfrac{\partial \delta p_k}{\partial p_k} = (\dfrac{\partial P_k}{\partial p_k} - 1)$ is already an infinitesimal, so that $\epsilon \dfrac{\partial \delta p_k}{\partial p_k}$ is a second-order infinitesimal, hence neglected. In words, since P_k and p_k differ by an infinitesimal, derivatives relative to them will differ on the same order. As a result, we obtain for the variations in the canonical variables,

$$\delta q_k = \epsilon \frac{\partial G}{\partial p_k}, \quad \delta p_k = -\epsilon \frac{\partial G}{\partial q_k}. \tag{3.112}$$

In terms of Poisson's brackets,

$$\delta q_k = \epsilon [q_k, G], \quad \delta p_k = \epsilon [p_k, G]. \tag{3.113}$$

Despite being not consistent with what was previously stated for function F, G is traditionally called the generating function of the infinitesimal canonical transformation.

3.6.1   Temporal Evolution as an Active Canonical Transformation

The time evolution of the canonical variables is given by

$$\delta q_k = \dot{q}_k dt = dt [q_k, H], \quad \delta p_k = \dot{p}_k dt = dt [p_k, H], \tag{3.114}$$

so that, by comparison with Eqs. (3.113), we obtain

$$\epsilon = dt \quad e \quad G = H. \tag{3.115}$$

The transformed variables represent the temporal evolution of the original dynamic variables,

$$Q_k(t) = q_k(t) + \delta q_k(t) = q_k(t) + \dot{q}_k dt = q_k(t + dt), \tag{3.116}$$

$$P_k(t) = p_k(t) + \delta p_k(t) = p_k(t) + \dot{p}_k dt = p_k(t + dt). \tag{3.117}$$

An important conclusion is reached: *The Hamiltonian is the generator of the temporal evolution of mechanical systems.*

In words, the evolution of a mechanical system between instants t and $t + dt$ is an infinitesimal canonical transformation generated by the Hamiltonian. Consequently, the evolution of the system in a finite time interval will be a succession of such transformations, a feature that will be explored in the context of the Hamilton-Jacobi theory.

3.6.2 Infinitesimal Variation of a Dynamical Quantity

Given the previous results, it becomes interesting to investigate how a generic quantity $A(q, p, t)$ varies when subjected to the active canonical transformation generated by H. We have,

$$\delta A = A(q + \delta q, p + \delta p, t) - A(q, p, t) = \sum_{k=1}^{n} \left(\frac{\partial A}{\partial q_k}\delta q_k + \frac{\partial A}{\partial p_k}\delta p_k\right). \qquad (3.118)$$

The term $\dfrac{\partial A}{\partial t}dt$ is null, since δA refers to the changes in A under the transformation $(q, p) \to (Q, P)$, which does not involve explicit time variation. Using the expressions (3.112) for δq_k and δp_k, we arrive to

$$\delta A = \epsilon[A, G]. \qquad (3.119)$$

If $G = H$ and $\epsilon = dt$, we get

$$\delta A = \dot{A}dt - \frac{\partial A}{\partial t}dt. \qquad (3.120)$$

Example 3.10 Here we illustrate the property of the angular momentum of generating infinitesimal rotations. Instead of rotating counter-clockwise, for example, the position vector of a particle by an infinitesimal angle $d\theta$, we can rotate clockwise the coordinate axes, see Fig. 2.6 in Sect. 2.6.

Taking arbitrarily the rotation axis as z, the matrix of clockwise rotation of axes is

$$\begin{bmatrix} cos(d\theta) & -sen(d\theta) \\ sen(d\theta) & cos(d\theta) \end{bmatrix} = \begin{bmatrix} 1 & -d\theta \\ d\theta & 1 \end{bmatrix}. \qquad (3.121)$$

The rotated (transformed) coordinates are given by

$$\begin{bmatrix} X \\ Y \end{bmatrix} = \begin{bmatrix} 1 & -d\theta \\ d\theta & 1 \end{bmatrix} \begin{bmatrix} x \\ y \end{bmatrix}, \qquad (3.122)$$

so that

$$\delta x = X - x = -yd\theta, \quad \delta y = Y - y = xd\theta, \quad \delta Z = Z - z = 0. \qquad (3.123)$$

Comparing these equations with the change of a quantity under an active canonical transformation, Eq. (3.119), and taking into account that the angular momentum of the particle is $L_z = xp_y - yp_x$, we find that

$$\delta x = d\theta\,[x, L_z], \quad \delta y = d\theta\,[y, L_z], \quad \delta z = d\theta\,[z, L_z]. \tag{3.124}$$

Indeed, $[x, L_z] = [x, xp_y - yp_x] = -y$ etc. This result establishes, within the formalism of Poisson brackets, that angular momentum is the generating function of infinitesimal rotations.

3.6.3　Poisson Brackets and Constants of Motion

Considering now the invariance of the Poisson brackets under canonical transformations and in view of Eq. (3.86), we see that for a quantity that is not explicitly dependent of t, $F(q, p)$,

$$\frac{dF}{dt} = [F, H]. \tag{3.125}$$

We draw important conclusions from this result:

(i) Every quantity that has zero Poisson brackets with H is a constant of motion, unless it is explicitly time-dependent.
(ii) Every constant of motion necessarily has null Poisson brackets with H.
(iii) Using Eq. (3.119) with $A = H$ and assuming that the generating function of the infinitesimal transformation G does not depend on time, we get

$$\delta H = \epsilon[H, G] = -\epsilon\frac{dG}{dt}, \tag{3.126}$$

from which we conclude that if $\delta H = 0$, meaning that H is invariant under an infinitesimal canonical transformation, the generator function G is a constant of motion. Conversely, constants of motion generate infinitesimal canonical transformations that keep H invariant.

3.7　Hamilton-Jacobi's Theory

We have already considered, in Sect. 3.5, the possibility of using canonical transformations so to facilitate the solution of Hamilton's equations. In the examples, however, the transformations that produce simplified Hamiltonians were obtained by inspection. We want now to systematize the procedure, obtaining a differential equation for the generating function.

Two possibilities are the most interesting. (i) In the first one, all transformed canonical variables are constant, which leads the new Hamiltonian to be zero, $K = 0$. (ii) The second corresponds to the transformed coordinates Q_k being all cyclic so that $K(P) = H(P)$ is also constant. In both cases, the integration of Hamilton's equations in the new variables becomes absolutely trivial. We thus focus on finding and solving a differential equation for the generator function that produces the desired transformation.

3.7.1 Case K = 0

In this case, Hamilton's equations become

$$\dot{Q}_k = \frac{\partial K}{\partial P_k} = 0 \Rightarrow Q_k = \beta_k, \quad \dot{P}_k = -\frac{\partial K}{\partial Q_k} = 0 \Rightarrow P_k = \alpha_k, \tag{3.127}$$

with α_k and β_k being constants.

A doubt can immediately come to student's mind: How can coordinates and momenta be constants? Well, they are not really "coordinates and momenta" any more. An example will clarify this point:

Example 3.11 Consider the problem of the 1D harmonic oscillator, whose solution is

$$q = Asen(\omega t + \phi) \quad \text{and} \quad p = m\omega A cos(\omega t + \phi). \tag{3.128}$$

We know that the amplitude A and the phase angle ϕ are constants determined by the initial conditions. We can propose as the variable Q,

$$Q = q^2 + \frac{p^2}{m^2\omega^2} = A^2, \tag{3.129}$$

that is, Q equals the square of the amplitude A, which is a constant fixed by the initial conditions. As the dynamics follows, the sum of the right-hand-side of the above equation keeps constant.

If Q informs the amplitude, P must be proportional to the phase angle $\phi = tan^{-1}(m\omega\frac{q}{p}) - \omega t$.

Exercise In the above example, show using the condition $[Q, P] = 1$, that $P = -\frac{m\omega}{2}\phi = -\frac{m\omega}{2}tan^{-1}(m\omega\frac{q}{p}) + \frac{m\omega^2}{2}t$, is also constant.

Not even the apparent explicit dependence on t here changes things, the terms defining P will sum up to a constant value. We see that there is no problem in taking Q_k and P_k as constants, as long as their knowledge provide the solution to the dynamic problem. The attentive reader should have already concluded that the constant values of Q_k and P_k will be proportional to the constants of motion, which are, in turn, written in terms of the initial conditions.

Back to the theory, according to Eq. (3.79), we can write,

$$0 = H(q, p, t) + \frac{\partial \mathbf{S}}{\partial t}, \tag{3.130}$$

where the reason for choosing the symbol $\mathbf{S}$ for the generating function, now called *Hamilton's principal function*, will become clear soon. The idea is to treat Eq. (3.130) as a differential equation whose solution is the generating function $\mathbf{S}$, that is, to build an operational method for this purpose, in place of inspection procedures considered so far. But this equation is not yet ready to use since, for that, H and $\mathbf{S}$ should depend on the same arguments. This is accomplished by choosing the form F_2 for the generating function $\mathbf{S} = \mathbf{S}(q, P)$. Being $P_k = \alpha_k$ constant,

$$\mathbf{S} = \mathbf{S}(q, \alpha, t), \text{ while, } p_k = \frac{\partial \mathbf{S}}{\partial q_k} = p_k(q, \alpha, t), \tag{3.131}$$

that is, the p_k also become (though still unknown) functions of q_k and t, exclusively. In other words, working with the variables q_k (and the time t, remembering that in the case $K \neq H$, $\frac{\partial \mathbf{S}}{\partial t} \neq 0$) we now obtain a differential equation for the generating function $\mathbf{S}(q, \alpha, t)$ in an operational form,

$$H(q_1, \ldots q_n, \frac{\partial \mathbf{S}}{\partial q_1}, \ldots \frac{\partial \mathbf{S}}{\partial q_n}, t) + \frac{\partial \mathbf{S}}{\partial t} = 0, \tag{3.132}$$

the famous *Hamilton-Jacobi equation*. A so-called *complete solution* of this equation, meaning a particular solution, dependent on n non-additive constants (additive constants can always be neglected, as only derivatives of $\mathbf{S}$ are involved, not $\mathbf{S}$ itself), will be the generating function of a canonical transformation that produces $K = 0$.

The turn back to the original variables is made with (see Table 3.1),

$$\beta_k = \frac{\partial \mathbf{S}}{\partial \alpha_k}, \tag{3.133}$$

from which we get $q_k(t)$, in the same way that Eq. (3.131) above gives $p_k(t)$.

Let us now take the time derivative of $\mathbf{S}(q, \alpha, t)$,

$$\frac{d\mathbf{S}}{dt} = \sum_{k=1}^{n} \frac{\partial \mathbf{S}}{\partial q_k} \dot{q}_k + \frac{\partial \mathbf{S}}{\partial t}. \tag{3.134}$$

Being $p_k = \frac{\partial \mathbf{S}}{\partial q_k}$ and, from (3.130), $\frac{\partial \mathbf{S}}{\partial t} = -H$ we can write,

$$\frac{d\mathbf{S}}{dt} = \sum_{k=1}^{n} p_k \dot{q}_k - H = L. \tag{3.135}$$

This result shows that the action $S = \int L \, dt$, if written in terms of q, α, t, differs from $\mathbf{S}$ at most by an addictive constant, which explains the choice of the symbol $\mathbf{S}$ for Hamilton's principal function. In words: *The action S, written in terms of coordinates and time, is the generating function $\mathbf{S}$ of the canonical transformation that produces $K = 0$ or, equivalently, transforms the dynamic variables into constants.*

As we have seen, the said constants are written in terms of the initial conditions $q_k(0)$ and $p_k(0)$. In practical terms, identifying the action playing such a role is useless, because to construct it as a function of q and t, we would have to know the solution of the problem under study. What must be done is to solve Hamilton-Jacobi's equation (3.132) to get $\mathbf{S} = S(q, \alpha, t)$ and then, obtaining $q_k(t)$ and $p_k(t)$ through Eqs. (3.131) and (3.133).

3.7.2 Separation of Cyclic Variables

According to the definition of H, Eq. (3.5), a cyclic coordinate in L is also cyclic in H. Suppose the coordinate q_n (for convenience) of a Hamiltonian is cyclic. In this case we know that p_n is already constant, so it can be identified as $p_n = P_n = \alpha_n$. In view of Eq. (3.131), $p_k = \dfrac{\partial S}{\partial q_k}$, S can be written

$$S = \alpha_n q_n + S'(q_1, ..., q_{n-1}, \frac{\partial S}{\partial q_1}, ..., \frac{\partial S}{\partial q_{n-1}}, \alpha_n, t). \tag{3.136}$$

This procedure can be repeated for as many cyclic coordinates as the Hamiltonian embodies. For example, in the case of a particle in a central field, $L = \dfrac{m}{2}(\dot{r}^2 + r^2\dot{\varphi}^2) - V(r)$ (see Eq. (2.89)), φ is cyclic, so that $p_\varphi = P_\varphi = \alpha_\varphi$ and $S = \varphi\alpha_\varphi + S'(r, \dfrac{\partial S}{\partial r}, \alpha_\varphi, t)$.

Exercise Obtain the Hamiltonian for the above case and show that φ is really cyclic.

3.7.3 Case K(=H) Constant, Separation of Time

If $\dfrac{\partial H}{\partial t} = 0$, the Hamilton-Jacobi equation admits the separation between coordinates and time in the form,

$$S(q_k, \alpha_k, t) = W(q_k, \alpha_k) - \alpha_1 t. \tag{3.137}$$

Taking this expression into Eq. (3.132) results in

$$H(q_k, \frac{\partial W}{\partial q_k}) = \alpha_1. \tag{3.138}$$

Considering $W(q, \alpha)$, called *Hamilton characteristic function*, as the new generating function, which generates a canonical transformation that leads to $K = H$, the equation above is a modified form of the Hamilton-Jacobi equation. According to Eq. (3.79),

$$K = H + \frac{\partial W}{\partial t} = H = \alpha_1. \tag{3.139}$$

This case corresponds to the second of the two discussed in the introduction of this section, whereby all coordinates Q_k are cyclic. In fact, using one of Hamilton's equations we have, $\dot{P}_k = 0 = -\frac{\partial K}{\partial Q_k}$, which can only be satisfied if Q_k is cyclic. All momenta P_k are constant and $K = H(P)$ (since $\frac{\partial S}{\partial t} = 0$) also results constant, equal to α_1.

This conclusion means that, for a conservative system, there are at least n independent constants of motion (it can be shown that the maximum number of constants is $2n$).

Note that the condition $K \neq 0$ is not in contradiction with Hamilton-Jacobi's theory. If we take the generating function as $S = W(q_k, \alpha_k) - \alpha_1 t$, the transformed Hamiltonian will be $K = 0$. If the generating function is W, then $K = H = \alpha_1$.

From Table 3.1, the other canonical variables are obtained from

$$p_k = \frac{\partial W}{\partial q_k}, \quad Q_k = \frac{\partial W}{\partial \alpha_k}. \tag{3.140}$$

Neglecting a harmless additive constant, we have $n - 1$ constants P_k which, added to $\alpha_1 = P_1$, form a set of n independent constants, determined according to the initial conditions of the problem. Any set of n constants satisfying the above stated, together with the variables Q_k obtained from Eq. (3.140), constitutes a set of canonical variables.

Example 3.12 The 1D harmonic oscillator provides a good example of how Hamilton-Jacobi's theory works. The Hamiltonian is $H = \frac{p^2}{2m} + \frac{m\omega^2 q^2}{2}$ so that Hamilton-Jacobi equation becomes,

$$\frac{1}{2m}(\frac{\partial S}{\partial q})^2 + \frac{m\omega^2 q^2}{2} + \frac{\partial S}{\partial t} = 0. \tag{3.141}$$

Taking $S = W(q) - \alpha_1 t$, the equation for $W(q)$ is

$$\frac{1}{2m}(\frac{\partial W}{\partial q})^2 + \frac{m\omega^2 q^2}{2} = \alpha_1, \tag{3.142}$$

from which $W = \int dq \sqrt{2m\alpha_1 - m^2\omega^2 q^2}$, so that

$$S = W(r) - \alpha_1 t = \int dq \sqrt{2m\alpha_1 - m^2\omega^2 q^2} - \alpha_1 t. \tag{3.143}$$

We don't have to perform the integration at this point, because

$$\beta = \frac{\partial S}{\partial \alpha} = \int \frac{m dq}{\sqrt{2m\alpha_1 - m^2\omega^2 q^2}} dq - t, \tag{3.144}$$

being this integral easier. Solving it, we finally find the known solution

$$q(t) = \sqrt{\frac{2\alpha_1}{m\omega^2}} sen(\omega(t + \beta)). \tag{3.145}$$

Example 3.13 An electron in a benzene molecule can be seen as suffering a potential like that of Hartmann [7], in spherical coordinates, $V(r, \theta) = \dfrac{A}{r} + \dfrac{B}{r^2 sen^2\theta}$, where A, B are constants. Considering the correspondent one-particle classical problem as a more practical application of Hamilton-Jacobi's theory, the Hamiltonian becomes

$$H = \frac{1}{2m}(p_r^2 + \frac{p_\theta^2}{r^2} + \frac{p_\phi^2}{r^2 sen^2\theta}) + \frac{A}{r} + \frac{B}{r^2 sen^2\theta}, \tag{3.146}$$

so that Hamilton-Jacobi's equation is,

$$\frac{1}{2m}[(\frac{\partial S}{\partial r})^2 + \frac{(\frac{\partial S}{\partial \theta})^2}{r^2} + \frac{(\frac{\partial S}{\partial \phi})^2}{r^2 sen^2\theta}] + \frac{A}{r} + \frac{B}{r^2 sen^2\theta} + \frac{\partial S}{\partial t} = 0. \tag{3.147}$$

Following Eq. (3.137) and noting that ϕ is cyclic, we write the generating function as,

$$S = \alpha_\phi \phi + W(r, \theta) - \alpha_1 t, \tag{3.148}$$

so that,

$$\frac{1}{2m}[(\frac{\partial W}{\partial r})^2 + \frac{(\frac{\partial W}{\partial \theta})^2}{r^2} + \frac{(\alpha_\phi)^2}{r^2 sen^2\theta} + \frac{A}{r} + \frac{B}{r^2 sen^2\theta}] = \alpha_1 \tag{3.149}$$

So far we did the standard procedure. Now, to solve Eq. (3.149), we try a separation of variables like $W(r, \theta) = W_r(r) + W_\theta(\theta)$ into it and multiply the resulting equation by r^2 obtaining, finally, the expected separation,

$$\frac{1}{2m}\left(\frac{dW_r}{dr}\right)^2 + A(r) - \alpha_1 = -\frac{\alpha_\theta^2 r^2}{2m} \quad ,$$

$$\frac{1}{2m}\left(\frac{dW_\theta}{dr}\right)^2 + \frac{\alpha_\phi^2}{2m\,sen^2\theta} + \frac{B}{r^2 sen^2\theta} = \frac{\alpha_\theta^2}{2m}. \tag{3.150}$$

Now, these equations are ready to be solved by quadrature.

Exercise Write out S, β_r, β_θ, and β_ϕ in terms of quadratures for the example above. Note that, to find the betas, there is no need to integrate previously the expression for S, as in the previous example.

Exercise Work out completely Hamilton-Jacobi's theory of a 3D projectile of mass m near Earth.

3.8   Hamilton-Jacobi's Perturbation Theory (HJ-PT)

The usefulness of Hamilton's theory can be greatly extended as we consider problems for which a perturbative approach is recommended. It is also particularly useful within Hamilton-Jacobi's version of Hamilton's equations. Let us consider $H_0(q, p, t)$ an unperturbed Hamiltonian for which a solution has already been found, by Hamilton-Jacobi's method, producing constant α and β with $K = 0$. Now, we consider an approximate solution for the Hamiltonian $H(q, p, t) = H_0 + \Delta H$, where ΔH is a small perturbation. The canonical transformation leading to K_0 can still be kept (see Sect. 3.5.4), but now the resulting α and β will no longer be constant, as well as the Kamiltonian will not be zero anymore. Instead,

$$K(\alpha, \beta, t) = H_0 + \Delta H + \frac{\partial S}{\partial t} = \Delta H(\alpha, \beta, t), \tag{3.151}$$

since $H_0 + \dfrac{\partial S}{\partial t} = K_0 = 0$.

Hamilton's equations will then become, on order n of perturbation theory,

$$\dot{\alpha}_{kn} = -\frac{\partial \Delta H(\alpha, \beta, t)}{\partial \beta_k}\bigg|_{n-1}, \quad \dot{\beta}_{kn} = \frac{\partial \Delta H(\alpha, \beta, t)}{\partial \alpha_k}\bigg|_{n-1}, \tag{3.152}$$

where $n = 1$ means that we are calculating the first-order corrections to α and β, by evaluating the derivatives with the unperturbed (zeroth-order) values α_0 and β_0, and so on. Higher-order corrections can be obtained using Eq. (3.152).

Example 3.14 Consider the 1D relativistic harmonic oscillator, with Hamiltonian given by, $H = \sqrt{m^2c^4 + p^2c^2} + \frac{1}{2}m\omega^2q^2$. We want to evaluate the relativistic effect on its motion for small but significant velocities, so that the relativist effect can be taken as a perturbation.

For $p << mc$, the truncated Taylor series gives $\sqrt{m^2c^4 + p^2c^2} = mc^2 + \dfrac{p^2}{2m} - \dfrac{p^4}{8m^3c^2}$, so that the Hamiltonian can be written as

$$H = H_0 - \frac{p^4}{8m^3c^2}. \tag{3.153}$$

where $H_0 = \dfrac{p^2}{2m} + \frac{1}{2}m\omega^2q^2$ is the harmonic oscillator Hamiltonian. So, we have the relativistic oscillator Hamiltonian as that non-relativistic, perturbed by $\triangle H = -\dfrac{p^4}{8m^3c^2}$.

In order to proceed to calculate the first-order correction to $q(t)$, we take advance of the solution Hamilton-Jacobi's equation for H_0, already made in Example 3.12, see Eq. (3.145). We easily obtain

$$p = \frac{\partial W}{\partial q} = \sqrt{2m\alpha}\cos(\omega(t + \beta)). \tag{3.154}$$

In terms of α and β, $\triangledown H$ becomes,

$$\triangledown H = -\frac{\alpha^2}{2mc^2}cos^4(\omega(t + \beta)). \tag{3.155}$$

The first-order corrections to α and β are given by Eq. (3.152). We have,

$$\dot{\alpha}_1 = -\frac{2\alpha^2\omega}{mc^2}cos^3(\omega(t + \beta))sen(\omega(t + \beta)). \tag{3.156}$$

Interestingly, the integral of this expression in a period of 2π vanish. This means that, in average, it has no net effect on the oscillator. In turn, we have for β,

$$\dot{\beta}_1 = \frac{\alpha}{mc^2}cos^4(\omega(t + \beta)). \tag{3.157}$$

There is no point here in going forward and taking the integral of this expression into $q(t)$, which would be dreadful, but we can see that its average is finite, so that a change is imposed to the phase of the oscillation.

Exercise A particle, previously free, begins to suffer a very tinny constant force F. Show that first-order Hamilton-Jacobi's theory describes already its full movement. Try to explain this result.

3.8.1   Action-Angle Variables

There is a very important application that, by itself, justifies the consideration of the case $K = H$ (cyclic Q's and constant P's), introduced in Sect. 3.7.3. *Multi periodic systems* are those for which each canonical pair (q_k, p_k) produces a *closed* (e.g., oscillators) or *periodic* trajectory in intervals Δq_k (e.g., pendulum) in the phase space. In cases the characteristic function is separable for such a system, in the sense that

$$W(q, \alpha) = \sum_{k=1}^{n} W_k(q_k, \alpha), \tag{3.158}$$

then

$$p_k = \frac{\partial W}{\partial q_k} = \frac{\partial W_k}{\partial q_k} = f_k(q_k, \alpha_1, ..., \alpha_n). \tag{3.159}$$

These trajectories f_k are projections on the plane (p_k, q_k) of the global trajectory of the system in phase space (q, p).

The *action variable J_k* is defined as the integral *over a complete revolution*,

$$J_k(\alpha) = \frac{1}{2\pi} \oint p_k dq_k. \tag{3.160}$$

and is constant, because it depends only the constants α_l. The relationship between them can be reversed,

$$\alpha_l = \alpha_l(J), \tag{3.161}$$

so the J_k can replace the α_k as the transformed momenta. Being $W(q, J)$ the generating function of the canonical transformation, the Kamiltonian will be the Hamiltonian written in terms of *action variables J*,

$$K = \alpha_1(J) = H(J). \tag{3.162}$$

The conjugate coordinates, the so called *angle variables*, will be, according to the second of equations (3.140),

$$\varphi_k = \frac{\partial W}{\partial J_k}. \tag{3.163}$$

Hamilton's equation,

$$\dot{\varphi}_k = \frac{\partial K}{\partial J_k} = \omega_k, \tag{3.164}$$

gives the fundamental frequencies ω_k of the system. So, whenever a multi-periodic system is exactly or approximately separable, the *fundamental frequencies* can be obtained without solving Hamilton's equations. Instead of doing a tedious general proof of this statement, lets resort to an example.

Example 3.15 The 2D harmonic oscillator is quite convenient for illustrating the action-angle method. Since its Hamiltonian is written as

$$H = \frac{p_x^2}{2m} + \frac{k_x}{2}x^2 + \frac{p_y^2}{2m} + \frac{k_y}{2}y^2 = \alpha_1, \tag{3.165}$$

The Hamilton-Jacobi's equation is separable once we take

$$W(x, y, \alpha) = W_x(x, \alpha) + W_y(y, \alpha), \tag{3.166}$$

leading to

$$\frac{1}{2m}\left(\frac{dW_x}{dx}\right)^2 + \frac{k_x}{2}x^2 = \alpha_x, \quad \frac{1}{2m}\left(\frac{dW_y}{dy}\right)^2 + \frac{k_y}{2}y^2 = \alpha_y, \quad \alpha_x + \alpha_y = \alpha_1. \tag{3.167}$$

As the trajectories in phase space are ellipses, the action variables are simply the areas of the ellipses, according to the definition equation (3.160). For the pair (x, p_x) we have $p_x = \dfrac{dW_x}{dx}$ and

$$\frac{x^2}{a^2} + \frac{p_x^2}{b^2} = \frac{x^2}{\left(\dfrac{2\alpha_x}{k_x}\right)} + \frac{p_x^2}{(2m\alpha_x)} = 1, \quad J_x = \frac{\pi ab}{2\pi} = \sqrt{\frac{m}{k_x}}\alpha_x, \tag{3.168}$$

with an analogous expression for the pair (y, p_y), so that,

$$K = \alpha_x + \alpha_y = \sqrt{\frac{k_x}{m}}J_x + \sqrt{\frac{k_y}{m}}J_y, \tag{3.169}$$

which exemplifies the situation expressed by Eq. (3.162). The characteristic frequencies are,

$$\omega_x = \frac{\partial K}{\partial J_x} = \sqrt{\frac{k_x}{m}}, \quad \omega_y = \frac{\partial K}{\partial J_y} = \sqrt{\frac{k_y}{m}}, \tag{3.170}$$

as expected. Note that for a 1D oscillator we would have

$$J = \sqrt{\frac{m}{k}}\alpha = \frac{E}{\omega}, \tag{3.171}$$

an expression that will be used forwardly.

However, the previous knowledge of $J(\alpha)$ is not common. In general, this dependence will be obtained by integration, represented in Eq. (3.160).

Exercise Obtain the characteristic frequencies of the 2D oscillator, Eqs. (3.170), by performing the integration of, $p_x^{\pm} = \pm\sqrt{2m\alpha_x - mk_x x^2}$ between the limits $x^{\pm} = \pm\sqrt{2\alpha_x/k_x}$, over a complete cycle.

3.9 Special Topic: A Classical Version of a Feshbach Resonance

Hamilton's equations and action-angle variables are still important research techniques. To exemplify this point, consider a three-particle system with two degrees of freedom, composed by an oscillator of two masses M separated by R and a third and much lighter mass m interacting with it at distance r from the $M - M$ center of mass, as seen in Fig. 3.4. This is a simplified version of a model representation of a positron interacting with a diatomic molecule [8]. Then (r, R) are our generalized coordinates, with associated momenta (p, P), respectively.

The 2D three-body potential energy, which describes the interaction between m and a two-body oscillator $M - M$, is written as the sum,

$$V(R, r) = V_{osc}(R) + V_{pos}(R, r). \tag{3.172}$$

The *osc* component in Eq. (3.172) is approximated by a harmonic oscillator potential.

$$V_{osc}(R) = \frac{1}{2}\mu\omega^2 R^2, \tag{3.173}$$

with reduced mass $\mu = 918.07$ and natural frequency $\omega = 0.02$ (in atomic units). Actually, the reader only need to know that Eq. (3.173), with these values for μ and ω, represents a harmonic potential for a hydrogen molecule, with R referred to the

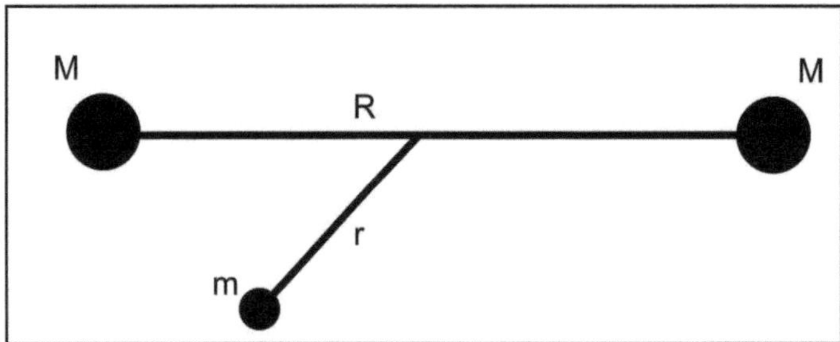

Fig. 3.4 The three-body problem with two heavy and a light particles

equilibrium bond distance of H_2. Also unnecessary here is the detailed form of the positron-molecule potential, but may be interesting to know that part of it is given in analytical form [9] as,

$$V_{cp}(R, r) = -\frac{\alpha_0(R)}{2r^4} f_\rho(r), \qquad (3.174)$$

where f damps the potential at a distance around and below $r = \rho$, and $\alpha_0(R)$ is the isotropic polarizability of the oscillator as a function of R, since the polarizability of a molecule varies with the distances between the atoms. The value used for ρ was 1.85 atomic units.

Hamilton's equations were solved numerically, starting from initial values of the variables (r, p, R, P). Figure 3.5 shows the positron trapped in a resonance state of the 3-body system, performing a number of oscillations and eventually escaping. Some trajectories are very sensible to initial conditions, interestingly displaying a chaotic nature.

The behavior of the oscillator interacting with the positron can be studied with the aid of the action variables J, ϕ, defined as $J = E_{osc}/\omega$ and $\phi = tg^{-1}\frac{P}{\omega\mu R}$. Figure 3.5 shows the final J as function of ϕ (horizontal dashed lines correspond to some chosen initial J values). The continuous parts of the curves correspond to collisions in which the oscillator mostly transfers some energy to the positron. Around $\phi = \pi$ the collisions are practically elastic. Beyond that, the oscillator can

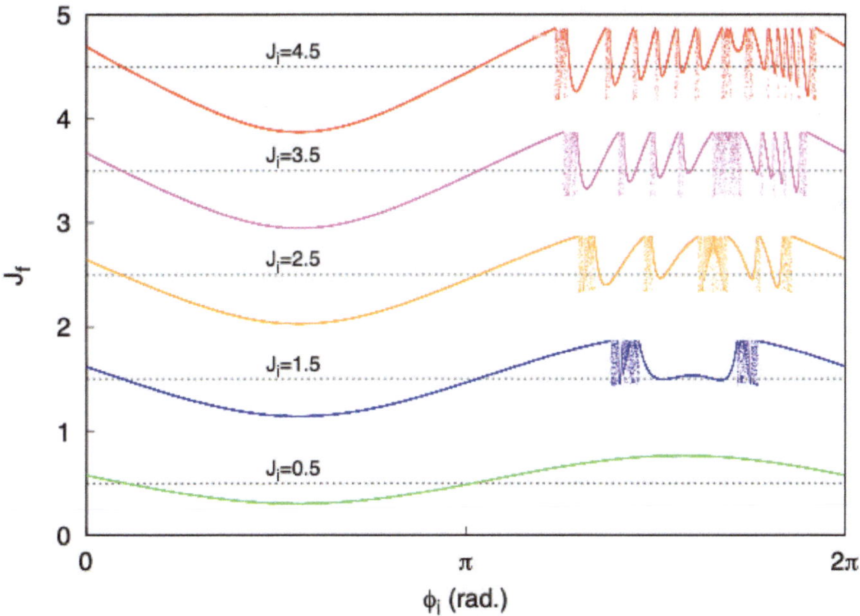

Fig. 3.5 Feshbach resonances

absorb energy of the positron, which, in turn, can get trapped in a resonance. These resonant states alternate between regular and chaotic behavior.

Action-angle variables keep being useful in research, mainly in astronomy and astrophysics, where obtaining characteristic frequencies of cosmic objects is fundamental.

3.10 Adiabatic Invariants

We have already seen several examples of invariant quantities and their theoretical relevance. Let us now introduce a kind of quasi-invariance in oscillating systems. They play an important role in some areas of physics, as astrophysics and plasma physics, for example. Adiabatic invariance was also a subject of discussions at the dawn of Quantum Mechanics.

If a parameter λ of the Hamiltonian of an oscillating system varies very slowly compared to its period of oscillation, some quantity involving λ can be identified as quasi-invariant, or *adiabatic invariant*. Being τ the period of oscillation, the slow character of λ is established by

$$\tau \frac{d\lambda}{dt} << \lambda. \tag{3.175}$$

Less rigorously, we can say that if ΔT is a characteristic time interval of the parametric variation, then,

$$\Delta T >> \tau. \tag{3.176}$$

The energy of such a system is no longer constant but, if the parameter changes slowly enough, we can assume that $\dot{E}$ is proportional to $\dot{\lambda}$, meaning that some relationship between E and λ holds approximately constant. As seen in the introduction to the chapter, this is a theme that brings the Hamiltonian theory closer to questions relative to contemporary physics, so we will develop it here with greater detail, mainly in what concerns its qualitative understanding.

Some hidden aspects of the concept are revealed as we consider a 1D oscillator, whose spring constant k varies very slowly in time [10]. The rate of change of its energy will be

$$\dot{E} = m\dot{x}\ddot{x} + kx\dot{x} + \frac{x^2}{2}\dot{k} = m\dot{x}(\ddot{x} + \frac{k}{m}x) + \frac{x^2}{2}\dot{k}. \tag{3.177}$$

We could resort to the common argument that the first two terms on the left, being fast and oscillatory, have zero temporal average. However, although the argument is valid for each term separately, it is not guaranteed that they go to zero at the same rate. What guarantees the correctness of taking the sum of the two terms as zero on average (for very small $\dot{k}$, the adiabatic limit) is the fact that the term in parentheses in Eq. (3.177) corresponds to the equation of motion for k constant, $\ddot{x} + \dfrac{k}{m}x = 0$, which must be obeyed at every value of k. In the following Example, we will see that what here appears to be a mere detail, is fundamental in other situations. The fast term $\dfrac{x^2}{2}$ can also be replaced by its time average $\dfrac{\langle V \rangle}{k}$, yielding,

$$\frac{dE}{dt} \simeq \frac{\langle V \rangle}{k}\frac{dk}{dt} = \frac{E}{2k}\frac{dk}{dt} \quad \Rightarrow \quad \frac{dE}{E} \simeq \frac{dk}{2k}, \tag{3.178}$$

in which we use the well-known result for the linear oscillator, $\langle V \rangle = \dfrac{E}{2}$. Assuming the equality in Eq. (3.178) and solving the resulting differential equation, we get

$$\frac{d}{dt}(\frac{E}{\sqrt{k}}) = 0 \quad \text{or} \quad \frac{E}{\omega} \quad \Rightarrow \quad \text{adiabatic invariant.} \tag{3.179}$$

All linear oscillators produce the same result [11].

However, there is no reason for the quantity $\dfrac{E}{\omega}$ being exactly constant for varying ω. What is then the real physical meaning of adiabatic invariance? To answer this question, let us consider an instantaneous change $\delta\omega$ in the frequency of the oscillator, which results in a change δE of its energy,

$$\delta\omega \quad \Rightarrow \quad \delta E = mx^2\omega\delta\omega. \tag{3.180}$$

This relation is justified by noting that only the potential energy depends directly on ω. Evaluating the ratio $\dfrac{E + \delta E}{\omega + \delta\omega}$, we get

$$\frac{E + \delta E}{\omega + \delta\omega} = \frac{E + \delta E}{\omega}(1 + \frac{\delta\omega}{\omega})^{-1} \simeq \frac{E + \delta E}{\omega}(1 - \frac{\delta\omega}{\omega}), \tag{3.181}$$

where the last step is due to the fact that $\dfrac{\delta\omega}{\omega} << 1$. Expanding the right-hand-side, we get

$$\frac{E}{\omega} + \frac{\delta E}{\omega} - \frac{E\delta\omega}{\omega^2} - \frac{\delta E\delta\omega}{\omega^2}. \tag{3.182}$$

Using Eq. (3.180) above, we get

$$\frac{\delta E}{\omega} = \frac{mx^2\omega^2\delta\omega}{\omega^2} = \frac{2V\delta\omega}{\omega^2}. \tag{3.183}$$

In a large number of oscillator cycles, replacing $2V$ by its time average $\langle 2V \rangle = E$, we see that the second and third terms of Eq. (3.182) cancel out, finally leading to

$$\frac{E + \delta E}{\omega + \delta\omega} = \frac{E}{\omega} + O\left(\delta E \delta\omega\right). \tag{3.184}$$

Now we can understand better the concept of adiabatic invariance: Infinitesimal first-order changes in ω and, consequently, in E, are accompanied by a second-order (product of infinitesimals) change in the ratio $\dfrac{E}{\omega}$, the adiabatic invariant. In other words, *the time derivative of an adiabatic invariant is approximately zero, within second-order infinitesimal deviation.*

In the sequence, we will state that the adiabatic invariants are the action variables J_k. It is to stand out that the calculation of J_k by Eq. (3.160) ignores any variation, of adiabatic nature or not, of any parameter, that is, the action variables seem "predestined" to be adiabatic invariants. Accordingly, in the case of the harmonic oscillator we have already seen, Eq. (3.179), that $J = \dfrac{E}{\omega}$, which coincides with the adiabatic invariant obtained above in another way. Thus, *if a parameter of the Hamiltonian of an oscillating system varies slowly, its first-order variations are accompanied by $\dot{J}_k = 0$; the J_k are adiabatic invariants.*

Another common doubt is: Is J the only adiabatic invariant in the harmonic oscillator problem? The question seems pertinent because we can also consider a slow variation of the mass of the oscillator, which yields

$$\dot{E} = m\dot{x}\ddot{x} + kx\dot{x} + \frac{\dot{x}^2}{2}\dot{m} = \dot{x}(m\ddot{x} + \dot{m}\dot{x} + kx) - \frac{\dot{x}^2}{2}\dot{m}. \tag{3.185}$$

The second step is necessary to input the correct equation of motion into the term in parentheses. As we can easily verify, contrary to the case of variable k, taking here $\langle m\dot{x}\ddot{x} + kx\dot{x} \rangle = 0$ (a common mistake!) would produce a wrong result. On the other hand, zeroing the term in parentheses of the equation above and taking the average of $-\dfrac{\dot{x}^2}{2}$, we get

$$\frac{dE}{dt} = -\frac{\langle T \rangle}{m}\frac{dm}{dt} \quad \text{ou} \quad \frac{dE}{E} = -\frac{dm}{2m}, \tag{3.186}$$

whose solution is

$$E\sqrt{m} \propto \frac{E}{\omega} \quad \Rightarrow \quad \text{adiabatic invariant}, \tag{3.187}$$

the same previous result.

Exercise Check this conclusion in detail.

Finally, we can ask what happens when both parameters, m and k, vary slowly and simultaneously. Remembering that the equation of motion remains $m\ddot{x} + \dot{m}\dot{x} + kx = 0$ and repeating the procedures for the independent variations of the parameters, we obtain

$$\frac{dE}{E} = \frac{1}{2}\left(\frac{dk}{k} - \frac{dm}{m}\right). \tag{3.188}$$

Now, since $\omega = \sqrt{\dfrac{k}{m}}$, we easily see that the right side of Eq. (3.188) is equal to $\dfrac{d\omega}{\omega}$, so that,

$$\frac{dE}{E} = \frac{d\omega}{\omega}, \tag{3.189}$$

which is equivalent to say again that $\dfrac{E}{\omega}$ is the adiabatic invariant.

The results show that the adiabatic invariant is the same, no matter which parameter undergoes slow variation. In fact, it was verified, for the harmonic oscillator at least, that the action variable is the adiabatic invariant, under slow changing of a generic parameter.

Example 3.16 Let us now consider a particle describing a circle of radius R in a frictionless horizontal plane. The centripetal force is the result of a constraint (a tensioned wire, for example), reducing the problem to one degree of freedom. The Lagrangian and the generalized momentum are, respectively,

$$L = \frac{m}{2}R^2\dot{\varphi}^2, \quad p_\varphi = mR^2\dot{\varphi}. \tag{3.190}$$

The coordinate φ is cyclic, so p_φ is constant. The action integral is then easily calculated as

$$J = \frac{1}{2\pi}\int p_\varphi d\varphi = \frac{p_\varphi}{2\pi}\int d\varphi = p_\varphi = l, \tag{3.191}$$

the generalized moment itself, where l is the modulus of the angular momentum of the particle. Considering, now, that R is slowly reduced (by pulling the string through a hole, for example), $p_\varphi = l$ will then be the adiabatic invariant.

This case is interesting because we can write [11] both E and ω as functions of l,

$$E = \frac{l^2}{mR^2} \quad \text{e} \quad \omega = \frac{l}{mR^2}. \tag{3.192}$$

The dependence of E and ω, on R and m is the same, so that the slow variation of either of the two parameters must keep the ratio $\dfrac{E}{\omega}$ constant, what was previously concluded for the harmonic oscillator.

Exercise Why doesn't the reasoning hold for fast changing parameters?

The general proof of the statement that the action variables are *the* adiabatic invariants[8] looks really scary for the beginner. Here we present a hopefully convincing argument for an 1D system, with no loss of generality. The Hamiltonian will be time-dependent through λ, that is, $H(q, p; \lambda)$. Taking its time derivative we have,

$$\frac{dH}{dt} = \frac{\partial H}{\partial q}\dot{q} + \frac{\partial H}{\partial p}\dot{p} + \frac{\partial H}{\partial \lambda}\dot{\lambda}. \tag{3.193}$$

The first two terms cancel each other as consequence of Hamilton's equations. So, we get,

$$\frac{dH}{dt} = \frac{\partial H}{\partial \lambda}\dot{\lambda}. \tag{3.194}$$

Keeping the same canonical transformation that makes $K = H$, even both depending slightly on t, we produce $H(J; \lambda)$. Taking the time derivative of H again, we get,

$$\frac{dH}{dt} = \frac{\partial H}{\partial J}\dot{J} + \frac{\partial H}{\partial \lambda}\dot{\lambda}. \tag{3.195}$$

The point is that, from Hamilton's equations, $\dfrac{\partial H}{\partial J} = \dot{\varphi}$, the frequency of the oscillator (see Eq. (3.164), a finite quantity that can be varying slowly or not. Comparing Eqs. (3.194) and (3.195), we see that $\dot{J}$ must be close enough to zero for them to agree to a good approximation, meaning an adiabatic invariant.

A general conclusion is that each system will have as many adiabatic invariants as its oscillating degrees of freedom.

Finally, we cannot forget the connection the concept of adiabatic invariants with physical properties of quantum systems. If a parameter external to the system (a field, for example) varies in such a way as to return to its initial value without, however, be sufficient to induce a change in the quantum state of the system, then some of its dynamic variables remain conserved, and should correspond, in the classical limit, to adiabatic invariants.

---

[8] See, for example, Goldstein's book in the bibliography.

3.11   A Transition to Quantum Mechanics: Canonical Quantization

Consider the Hamiltonian operator $\hat{A}$ of quantum mechanics. Its eigenvalue equation is

$$\hat{A}\psi(t) = a(t)\psi(t), \tag{3.196}$$

where the dependence of ψ on coordinates and spin and its labelling with quantum numbers are avoided just for simplicity. Schrödinger's equation, which determines the time evolution of ψ is written as

$$i\hbar\frac{\partial\psi}{\partial t} = \hat{H}\psi, \tag{3.197}$$

in which $\hat{H}$ is the Hamiltonian operator, of central importance in quantum theory. Taking the total time-derivative of (3.196) we obtain,

$$\frac{\partial\hat{A}}{\partial t}\psi + \hat{A}\frac{\partial\psi}{\partial t} = \frac{da}{dt}\psi + a\frac{\partial\psi}{\partial t}. \tag{3.198}$$

Now, using Eq. (3.197) for $\dfrac{\partial\psi}{\partial t}$ and realizing from Eq. (3.196) that $\hat{A}\hat{H}\psi = \hat{H}\hat{A}\psi = a\hat{H}\psi$, we get

$$\frac{\partial\hat{A}}{\partial t}\psi + \frac{i}{\hbar}[\hat{H},\hat{A}]\psi = \frac{da}{dt}\psi, \tag{3.199}$$

in which we used that $a = a(t)$ exclusively and recalled the definition of the quantum commutator $[\hat{A},\hat{B}] = \hat{A}\hat{B} - \hat{B}\hat{A}$.

On the other hand, Eq. (3.196) allows us to write $\dfrac{d}{dt}(a-\hat{A}) = 0$, then $\dfrac{da}{dt} = \dfrac{d\hat{A}}{dt}$, so getting finally

$$\frac{d\hat{A}}{dt} = \frac{i}{\hbar}[\hat{H},\hat{A}] + \frac{\partial\hat{A}}{\partial t} \tag{3.200}$$

The similarity of Eqs. (3.200) and (3.86) strikes the eyes, as long as we identify a connection between the Poisson bracket of two dynamical quantities with the commutator of their related operators. The quantization rule for replacement of classical relations by corresponding quantum relations,

$$[A,B] \longrightarrow -i\hbar[\hat{A},\hat{B}] \tag{3.201}$$

supports, for example, the well known quantum mechanics postulations $x \longrightarrow \hat{x}$ (meaning "multiply by x") and $p \longrightarrow \hat{p} = -i\hbar\dfrac{\partial}{\partial x}$.

Exercise Verify the proposition above.

In quantum mechanics, the concept of energy becomes central, so as to the operator $\hat{H}$, obtained from H by using the canonical quantization rules above, playing also a fundamental rule in the theory.

3.12 Final Considerations on Hamiltonian Mechanics

The Hamiltonian formulation is particularly useful in simulations of many-particle systems, where it explores the greater ease of dealing numerically with differential first-order equations in phase space. For the better understanding of this feature, it is important to realize that each of Hamilton's equations must be solved simultaneously to its pair. So, the substitution of one into the other is not a proper procedure, specially for teaching. Connection to statistical physics is made through Liouville's theorem.[9] But it is the central importance of the canonical (symmetrical) relation between coordinates and moment that makes Hamilton's dynamics so powerful, as shown in further developments as Hamilton-Jacobi's and Action-Angle theories. In turn, as viewed in the last section, Hamilton's formulation is absolutely central for developing the canonical quantization, one of the most successful ways to present quantum mechanics. Classical mechanics can be then seen as a limiting case of quantum mechanics.

---

[9] See Goldstein's book in the bibliography.

Chapter 4
Lagrangian Theory of Classical Fields

Abstract Instead of a regular introduction to the theme Classical Fields, this chapter can be seen as a continuation of Chap. 1, in the sense that classical field equations are obtained from guessing appropriate Lagrangians, with the help of imagination and symmetry properties of space-time, in order to use them in the context of Hamilton's principle. The need of including time as a new coordinate is discussed for relativistic and non-relativistic fields.

Keywords Lorentz invariants · Field Lagrangians · Hamilton's principle for fields · Static · Stationary and variable fields · Particle-field interactions

This chapter can be seen as a continuation of the first chapter, sharing the main driven motive of showing that once appropriate Lagrangian is guessed, Hamilton's principle allows developments much beyond Newtonian mechanics. So far we have treated motion of particles, now is the time to consider fields and ask whether the least action principle[1] works for them also. We will see that the answer not only is yes, but that the action principle is rather *the* way to treat fields.

4.1 Some Considerations Concerning Invariance Under Change of Inertial Frames

In Newtonian mechanics we deal easily with invariance of physical laws, showing first that these laws must be written in terms of scalars, vector and tensors, in order to be invariant under rotations and translations of reference frames. Then, we show that Newton's second law keeps its form under exchange of inertial reference frames using Galilean transformations. This is all we need!

---

[1] This term seems better now, since Hamilton's principle was developed for particles motion.

J. R. Mohallem, *Lagrangian and Hamiltonian Mechanics*,
https://doi.org/10.1007/978-3-031-55202-1_4

In the relativistic regime things are a little different. Physics equations must be invariant under Lorentz, instead of Galilean, transformations, which assures their invariance under rotations as well. A *Lorentz invariant* equation or formulation is such that time is treated on equal footing with space coordinates. This is usually done within Minkowiski's quadrivector (or, more generally, quadritensor) notation and the corresponding formalisms are called *covariant*. In fact, things are a little more involved, since an equation term, the vector potential of electrodynamics for example, can be not invariant, but the Maxwell equations involving it are still covariant. A *manifestly covariant* formulation basically means that the physical quantities and equations are written in tensor form so that they keep the same form in all inertial frame. This relatively confuse terminology does not mean that equations written in a non-covariant form are necessarily not invariant under Lorentz transformations.

For example, Maxwell's equations of electromagnetism, written in the usual way, in terms of vector fields are, as we already know, invariant under Lorentz's transformations, but time and coordinates are not treated in the covariant form. Tough being a difficult problem, their invariance can be verified in the vector form. So, Maxwell's equations are covariant under Lorentz's transformations and also manifestly covariant when written in the tensor form. Considering the scope of this book, it is not our interest to use the covariant (or manifestly covariant) formalism. Instead, we keep the scalar/vector form of equations that the student is used to. Anyway, these subtleties should not bother us here too much.

Now, consider the relativistic Lagrangian, Eq. (1.37), for a particle subject to a potential. Let us first turn off the potential so that only the "kinetic" term remains (a free particle). How was this Lagrangian guessed? It was not by magic, now we know. A first hint is that the relativistic L should reproduce the non-relativistic kinetic energy $T = \frac{1}{2}mv^2$, in the limit $v \ll c$, what it really does. Moreover, we have symmetry arguments. In the non-relativistic case we explored Galilean invariance of either L or S, since t is Galilean invariant. But in the relativistic regime t is not Lorentz invariant so that we need to demand the invariance of the action S, after all, a time integral. The action must be a scalar, not only a number but an invariant number under Lorentz transformations. So, hints for constructing L must come from known relativistic invariants.

The most known relativistic invariant is the *proper time* τ defined, in differential form, as

$$d\tau^2 = dt^2 - (dx^2 + dy^2 + dz^2)/c^2. \tag{4.1}$$

Exercise Use Lorentz transformations and limit τ to a 2D motion in space-time, represented by (t,x), just for simplicity (in general we have four-dimensions, one in time-type and three space-type) to prove its relativistic invariance. Moreover, prove that $dt^2 + dx^2$ is not Lorentz invariant.

Now, keeping the 2D restriction of the previous exercise,

$$d\tau = \sqrt{dt^2 - dx^2} = dt\sqrt{1 - (\frac{dx}{dt})^2} = dt\sqrt{1 - v^2}. \qquad (4.2)$$

Thus, the relativistic Lagrangian of a free particle becomes simply $-mc^2 d\tau$, which is a scalar. Realizing that the "volume" element $dt\,dx\,dy\,dz$ is known to also be Lorentz invariant, we conclude that the corresponding action, $S = -mc^2 \int d\tau$ is clearly invariant and manifestly covariant, since it has the same form in all inertial frame. The Lagrange equation for the free particle will then be Lorentz covariant.

On the other hand, a problem comes out when we consider a potential $V(\vec{r})$, which can be not frame-independent, even being a scalar in space. To be Lorentz invariant, it must be a scalar (a 0-rank tensor) in space-time, which is not guaranteed. The classical example is the pure Coulombic potential, which is not Lorentz invariant. In fact, while an observer in rest relative to a charge sees just the Coulomb potential, another inertial observer in motion will see a moving charge, so that the potential will no longer be purely Coulombic. Furthermore, it assumes instantaneous interactions, which violates relativity. In what follows we disregard these cases, considering just covariant potentials. The interested reader should enjoy the information that $V(\vec{r})$ is, in fact, a component of the quadri-vector potential.

We already had a taste of what comes forward when discussed generalized potentials in Sect. 1.7, now is time to go deeper on this matter. But first, some considerations on classical fields are in order.

4.2 Classical Fields

Consider an empty box whose wall temperatures are controlled. At first, all walls are at the same temperature T. Soon the thermal equilibrium will be established inside the box, so that all points will have the same temperature, say T_0, at any time. We say we have a *static* temperature field $T(x, y, z) = T_0$. Let us now suppose the temperature of the left wall is raised to $T_1 > T_0$. Heat will flow from left to right so that, when a steady state is reached, there will be a constant temperature gradient inside in the left-right x direction But at each point, T will still be constant, at least ideally. We now say we have a *stationary* temperature field $T(x, y, z) = T(x)$. Now, consider that all walls have their temperatures varying, say randomly. The temperature at each point inside the box will vary in *space and time*, $T = T(t, x, y, z)$. We say that we have a *variable temperature field* inside the box. This is a new quantity as compared to particles. Here, (t, x, y, z) are all independent parameters that fix the value of $T(t, x, y, z)$. Particularly, in all cases, (x, y, z) are no longer coordinates on a trajectory, but all points in space. The degree of freedom here is the field itself and t, x, y, z are the parameters that fixed it.

A *field* is a measurable quantity $\Phi(t, x, y, z)$ that is defined for each point in ordinary space at a given instant. The way time is related to the point coordinates in two inertial frames define the nature of the field, relativistic or not. Evidently, if t is frame-independent, the field is non-relativistic. If it is mixed to the coordinates in Lorentz transformations, the field is relativistic, as for example, the electromagnetic field, and we refer to points in *space-time*. In turn, the term "classical" means here "non-quantum".

The theme "classical fields" embodies a large range of systems, including continuous matter oscillations. On the other hand, the non-relativistic wave propagation in a material medium is a subject of basic physics, where the Lagrangian formulation is not necessarily helpful. Indeed, this chapter is devoted to cases for which there is no other appropriate tool to approach the physics of the systems, so that Hamilton's principle is the actual starting point of the theory, through the choice of appropriate Lagrangians, supported by symmetry considerations.

However, an application to a non-relativistic continuous system introduce some elements of field theory, being so useful for beginners. Consider the field $\Phi(x, t)$ of vertical dislocations of a continuous rope with fixed extreme points. Elementary transition from discrete points to the continuum[2] renders the *Lagrangian density* (see definition below) to the form,

$$\mathsf{L} = \frac{\lambda}{2}\left(\frac{\partial \Phi(x, t)}{\partial t}\right)^2 - \frac{\tau}{2}\left(\frac{\partial \Phi(x, t)}{\partial x}\right)^2, \tag{4.3}$$

in which λ is the linear mass density and τ is the tension on the rope. The reason the Lagrangian appears with a different symbol will be explained soon. Despite being deduced with the use of Newton's laws, a close look to this Lagrangian shows that it has not the $T - V$ form. Actually, there is no potential energy term yet in this Lagrangian. In the deduction of Eq. (4.3) the central object was not a generalized coordinate anymore, but the field amplitude Φ was the relevant degree of freedom, which depends on both t and x.

The two terms of this Lagrangian, in fact, resemble that of the kinetic energy of a 1D-particle if we consider the field ϕ as our "generalized coordinate". But, if this is the case, where does the second term come from? The clue is that the field $\Phi(x, t)$ varies continuously with both x and t; the same reason that induces us to include the immediate infinitesimal variation of Φ on t, now applies also for x!

In the next sections these (momentarily) weird features will be clarified. For now I just inform readers that this Lagrangian produces the expected free wave equation,

$$\frac{1}{v^2}\left(\frac{\partial^2 \Phi}{\partial t^2}\right) - \left(\frac{\partial^2 \Phi}{\partial x^2}\right) = 0, \tag{4.4}$$

where $v = \sqrt{\dfrac{\tau}{\lambda}}$ is the wave velocity.

---

[2] See Lemos' book in the bibliography.

A field can be *scalar* (as the temperature in a room), *vector*, as the gravitational or the electromagnetic fields, etc.[3] Using Φ to represent a generic field (or a component of a vector field), we now have to consider its own role and the roles of the variables (t, x, y, z) it depends on. The field does not describe a trajectory in space as t goes on but, instead, it is defined in space and time, or in space-time. So, from here on we start considering (t, x, y, z) as having the status of basic independent variables, having the same status of t in particle Lagrange theory.[4] However, to assert them an equal physical dimension, namely length, we multiply the time variable with a parameter α having the dimension of velocity, $t \to \alpha t$, instead. To save notation, however, we will write terms like $\partial(\alpha t)$ as simply $\partial \alpha t$.

Despite these features are common to non-relativistic fields, as exemplified with the wave-in-a-rope above, in formal terms (though not qualitatively) they put us, so to say, halfway to relativity. In fact, besides the variation either in space and time, if the field represents an interaction, its infinitesimal variation in space overcomes the Newtonian idea of "action at a distance". These features might explain some similarities between relativistic and non-relativistic fields.

The field variations in space and time are evaluated by the partial first-derivatives $\dfrac{\partial \Phi}{\partial \alpha t}, ..., \dfrac{\partial \Phi}{\partial x}$. So, according to what we have already learned for Lagrangians, for this generic field we write,

$$\mathsf{L} = \mathsf{L}(\Phi, \frac{\partial \Phi}{\partial t}, \frac{\partial \Phi}{\partial x}..., t, x, y, z), \tag{4.5}$$

that is, Φ plays an analogous role of $\vec{r}$ in Eq. (1.22), while the roles of x, y, z here become analogous to that of t there. We are defining here a 4-dimension field which is normally called a (3+1) field, where the label 1 corresponds to the temporal variable. Note, in consequence, that a particle might be referred to as a (0+1) field, since $\vec{r}(t)$ becomes the very field in this case. Only the t dimension applies and, once we identify $\vec{r}(t) \equiv \Phi(t)$, we can "give a spoiler" of what comes next by writing the corresponding Lagrange's particle-as-a-field equation as $\dfrac{d}{dt} \dfrac{\partial \mathsf{L}}{\partial (\frac{\partial \Phi}{\partial t})} - \dfrac{\partial \mathsf{L}}{\partial t} = 0.$

The action integral becomes

$$S = \int_a^b \mathsf{L}(\Phi, \frac{\partial \Phi}{\partial x}, ..., x, y, z, t)dxdydzdt, \tag{4.6}$$

---

[3] We can have *tensor* fields of any rank. The gravitational field is, in fact, a tensor field of rank 2, but this feature does not bother us here.

[4] Note the different status of x, y, z from particle Lagrangian theory, where they represent points in configuration space, while here they represent all points in space.

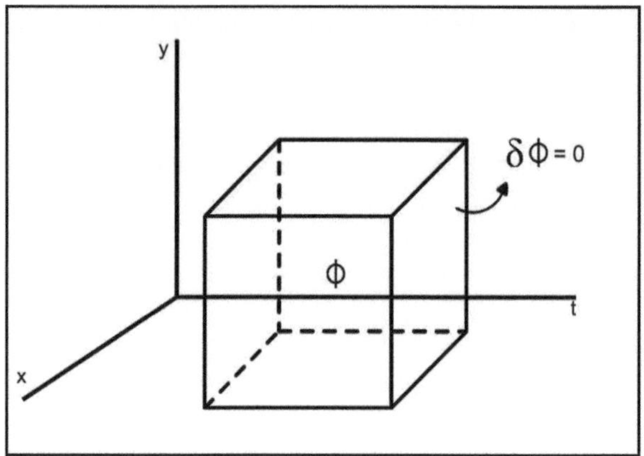

Fig. 4.1 Φ variation in the 3D space x, y, t. The faces of the cube are the locations where Φ is fixed, while it is varied inside the cube

where the integration limits a and b mean fixed "points" in the 4-dimensional space, that is, x, y, z and t are fixed in the extremes of the varied trajectories, as shown, unfortunately just for (2+1) dimensions, in Fig. 4.1.

In view of Eq. (4.6), it becomes clear that L should be called *Lagrangian density*, since, in order to keep the dimension of S, the dimension of L must be energy/volume. However, using the term *Lagrangian* also here is customary, but it seems convenient to use another symbol, L, as we started doing already.

In order to check the correctness of a guessing Lagrangian for Eq. (4.5), we need to impose $\delta S = 0$ and check the resulting Lagrange's equation, concerning its description of the field dynamics. Note that the field in consideration has just one degree of freedom, which is represented by Φ itself, so just one Lagrange's equation will be obtained. For performing the action variation we will restrict ourselves to a scalar bi-dimensional field. The generalization is straightforward.

4.3   Equations of Motion for Fields

Let us consider the (1+1) field $\Phi(t, x)$. The action integral is then

$$S = \int_a^b \mathsf{L}\left(\Phi, \frac{\partial \Phi}{\partial t}, \frac{\partial \Phi}{\partial x}, x, t\right) dx dt. \tag{4.7}$$

The functional dependence of L on its arguments is not relevant for now. Performing the variation $\delta S = 0$ under variation $\delta \Phi$ of Φ, in which the extremes $a = (t_1, x_1)$ and $b = (t_2, x_2)$ are held fixed as usual, and in which the variations are made with t and x fixed, it might be easy for the reader, at this point, to arrive to,

$$\int_{t_1}^{t_2} dt \int_{x_1}^{x_2} dx [\frac{\partial L}{\partial \Phi} - \frac{\partial}{\partial t} \frac{\partial L}{\partial (\frac{\partial \Phi}{\partial t})} - \frac{\partial}{\partial x} \frac{\partial L}{\partial (\frac{\partial \Phi}{\partial x})}] \delta \Phi = 0. \tag{4.8}$$

Exercise Check the statement above.

Considering the arbitrariness of $\delta \Phi$, we are lead to Lagrange's equation for the field Φ,

$$\frac{\partial}{\partial t} \frac{\partial L}{\partial (\frac{\partial \Phi}{\partial t})} + \frac{\partial}{\partial x} \frac{\partial L}{\partial (\frac{\partial \Phi}{\partial x})} - \frac{\partial L}{\partial \Phi} = 0. \tag{4.9}$$

Note that if we call $\dot{\Phi} \equiv \frac{\partial \Phi}{\partial t}$, this equation assumes a form which is very close to Lagrange's equation for a generalized coordinate of a particle, Eq. (1.21) or (2.29), except for the extra second term involving $\Phi' \equiv \frac{\partial \Phi}{\partial x}$. The previous procedure can be extended for (y, z) or more dimensions. A general expression of Lagrange's equations for scalar fields is then,

$$\sum_\nu \frac{\partial}{\partial x^\nu} \frac{\partial L}{\partial (\frac{\partial \Phi}{\partial x^\nu})} - \frac{\partial L}{\partial \Phi} = 0, \tag{4.10}$$

where $x^\nu = t, x_1, x_2 \ldots .$ The presence of a second field means one more degree of freedom, so there will be one more equation like (4.10) and so on.

4.4   Searching for Field Lagrangians

The credibility of these equations of motion, or in fact of each proposed Lagrangian, depends on confronting their solutions with the real behavior of the fields. We then move to discussing the forms of the Lagrangians in specific cases.

4.4.1   A Static Field

Consider the simplest time-independent scalar field, e.g., a temperature field in a room kept in thermal equilibrium, Φ_T. Considering the analogy with a free particle, and recalling that (x, y, x) are now parameters analogous to t in the "kinetic" term, a likely Lagrangean is,

$$L = \frac{1}{2}[(\frac{\partial \Phi_T}{\partial x})^2 + (\frac{\partial \Phi_T}{\partial y})^2 + (\frac{\partial \Phi_T}{\partial z})^2] \tag{4.11}$$

Using Eq. (4.10) we easily get

$$\frac{\partial^2 \Phi_T}{\partial x^2} + \frac{\partial^2 \Phi_T}{\partial y^2} + \frac{\partial^2 \Phi_T}{\partial z^2} \equiv \nabla^2 \Phi = 0. \tag{4.12}$$

This result means that the average temperature in the neighbourhood of a generic point (x, y, z) is equal to the temperature at that point, which can be seen as the very definition of thermal equilibrium. That was already known for sure but the example is valid to check for consistency of the theory.

4.4.2 A Relativistic Field

The notion of a field, particularly a relativistic one, is quite useful because it permits the substitution of the old concept of instantaneous interaction at distance by a new local, non-instantaneous interaction transported through the field itself to another position in space, which takes time. A relativistic field is then defined as one that satisfies Lorentz transformations, treating time on the same footing of coordinates (or vice-versa). As in Lagrangian theory of particles not obeying the $T - V$ prescription, here also, obtaining the Lagrangian for a non-relativistic field sometimes is something like an "arrival account", that is, we know the wave equation and search for the Lagrangian that fits it. Some clues help in this search: Lorentz invariance and other symmetries are fundamental. Besides, a *free field*, that is, in the absence of a potential $V(\Phi)$, must propagate with velocity equal to c. Actually this feature should be (and is!) an output of the theory, but there is no problem in using known results to shorten our work.

Our previous experience with Lagrange's theory might lead us to consider a Lagrangian to use in Eq. (4.9) with the form, $L = \frac{1}{2}[\frac{1}{\alpha^2}(\frac{\partial \Phi}{\partial t})^2 + (\frac{\partial \Phi}{\partial x})^2] - V(\Phi)$, where $V(\Phi)$ is some (density of) potential energy term, the *field potential*, and the derivative terms in x and t have the same signal. This is not in accordance with Eq. (4.3), but, after all, we want to treat time strictly on the same footing with the other coordinates. However, I can tell you, in advance, that this is not going to work, so there is no point in pursuing with it. The important point is to find why not.

Exercise Confirm the statement above.

A hint comes from relativity, where αt does not play a fully symmetric role to (x, y, z). For example, we have already seen that the invariant proper time is given by $\tau^2 = t^2 - (x^2 + y^2 + z^2)/c^2$. So, we surrender and, taking $\alpha = c$, propose the term in t of the Lagrangian with an inverse sign relative to the spacial terms. Then, L will have the same "structure" of Eq. (4.1), in what concerns its "kinetic" term. Recall that this choice should make the Lagrangian Lorentz invariant, provided $V(\Phi)$ also is.

$$\mathsf{L} = \frac{1}{2}[\frac{1}{c^2}(\frac{\partial \Phi}{\partial t})^2 - (\frac{\partial \Phi}{\partial x})^2] - V(\Phi). \tag{4.13}$$

Taking it to Lagrange's equation (4.9), after obtaining the derivatives $\dfrac{\partial L}{\partial(\frac{\partial \Phi}{\partial t})} = \dfrac{1}{c^2}\dfrac{\partial \Phi}{\partial t}$ and $\dfrac{\partial L}{\partial(\frac{\partial \Phi}{\partial x})} = -\dfrac{\partial \Phi}{\partial x}$, we obtain the wave equation

$$\frac{1}{c^2}\frac{\partial^2 \Phi}{\partial t^2} - \frac{\partial^2 \Phi}{\partial x^2} + \frac{\partial V(\Phi)}{\partial \Phi} = 0. \tag{4.14}$$

It represents a propagation of a wave in the x-direction with velocity c under the influence of a "force" $-\dfrac{\partial V(\Phi)}{\partial \Phi}$. The field Φ in this case is the wave amplitude.

The previous example for $V(\Phi) = 0$ corresponds to a field propagating with the velocity of light c in the vacuum.

In spite of all arguments, we should be also able to check if the Lagrangian (4.13) is really Lorentz invariant. This check can be done easy.

Exercise After understanding the next development, prove that Lagrangian (4.13) is Lorentz invariant.

A little challenger (though equivalent) is to check if the wave equation (4.14) is Lorentz invariant, so it is done here. Without loss, we can keep working with a (1+1) system in space-time. The Lorentz transformations are then,

$$x' = \gamma(x - ut) \ \text{ and } \ t' = \gamma(t - \frac{u}{c^2}x), \tag{4.15}$$

where $\gamma = \sqrt{1 - \frac{u^2}{c^2}}$, where u is the relative velocity of the reference frames.

In order to change the derivative terms of Eq. (4.14) to the transformed variables, we need the terms

$$\frac{\partial x}{\partial x'} = \gamma, \ \frac{\partial x'}{\partial t} = -\gamma u, \ \frac{\partial t'}{\partial x} = -\frac{\gamma u}{c^2} \ \text{and} \ \frac{\partial t'}{\partial t} = \gamma. \ \text{Thus,}$$

$$\frac{\partial \Phi}{\partial x} = \frac{\partial \Phi}{\partial x'}\frac{\partial x'}{\partial x} + \frac{\partial \Phi}{\partial t'}\frac{\partial t'}{\partial x} = \gamma\frac{\partial \Phi}{\partial x'} - \frac{\gamma u}{c^2}\frac{\partial \Phi}{\partial t'}.$$

For the second derivation we need to repeat carefully the above procedure, using again the chain rule as done above, for each term. The result is (check it!),

$$\frac{\partial^2 \Phi}{\partial x^2} = \gamma^2(\frac{\partial^2 \Phi}{\partial x'^2} + \frac{u^2}{c^4}\frac{\partial^2 \Phi}{\partial t'^2} - \frac{2u}{c^2}\frac{\partial^2 \Phi}{\partial x'\partial t'}).$$

The procedure to obtain $\dfrac{\partial^2 \Phi}{\partial t^2}$ is similar and yields,

$$\frac{\partial^2 \Phi}{\partial t^2} = \gamma^2 (u^2 \frac{\partial^2 \Phi}{\partial x'^2} + \frac{\partial^2 \Phi}{\partial t'^2} - 2u \frac{\partial^2 \Phi}{\partial x' \partial t'}).$$

Taking the two last results into Eq. (4.14) with $V(\Phi) = 0$, after some manipulations, we get

$$\frac{1}{c^2} \frac{\partial^2 \Phi}{\partial t'^2} - \frac{\partial^2 \Phi}{\partial x'^2} = 0.$$

That is, under Lorentz transforming (x, t) to (x', t'), the form of the wave equation did not change. It is Lorentz invariant!

Exercise Following the same steps above, show that the relativistic free wave equation is not invariant under a Galilean transformation.

Some considerations are due here. It is clear from the previous development that the presence of the term c^2 in the Lagrangian (4.13), as well as in the denominator of the wave equation (4.14,) is central to the demonstration. The student must be aware, however, that it is not a "simple detail". Material waves propagate with velocity v, see Eq. (4.4), relative to the material media, a particular reference frame. In turn, free relativistic waves propagate with velocity c relative to *all* inertial reference frames. There is a profound change of paradigm behind the swap from v to c in the wave equations. The formal similarity can be explained, again, by considering space and time parameters on a common footing in field theory, whether relativistic or not.

As we turn $V(\Phi)$ on, new possibilities come out. One of the most famous is the Klein-Gordon field, whose Lagrangian is,

$$\mathsf{L} = \frac{1}{2}[\frac{1}{c^2}(\frac{\partial \Phi}{\partial t})^2 - (\frac{\partial \Phi}{\partial x})^2 - (\frac{\partial \Phi}{\partial y})^2 - (\frac{\partial \Phi}{\partial z})^2] - \frac{\mu^2}{2}\Phi^2, \tag{4.16}$$

where μ is a constant.

Lagrange's equation becomes,

$$\frac{1}{c^2} \frac{\partial^2 \Phi}{\partial t^2} - \frac{\partial^2 \Phi}{\partial x^2} - \frac{\partial^2 \Phi}{\partial y^2} - \frac{\partial^2 \Phi}{\partial z^2} + \mu^2 \Phi = 0. \tag{4.17}$$

Lagrangian (4.16) resembles that of the harmonic oscillator. Ignoring, for while, the terms of partial derivatives with respect to space coordinates and multiplying the wave equation (4.17) by c^2 results in $\dfrac{\partial^2 \Phi}{\partial t^2} + c^2 \mu^2 \Phi = 0$. This harmonic-oscillator-like result induces us to consider the term multiplying Φ with something related to mass. In fact, the potential $V(\Phi)$ is said as *giving mass to the field*. In quantum field theory, the Klein-Gordon field applies to scalar mesons.

4.4.3 Particle-Field Interactions

We have already seen, in the first chapter, that the Lagrangian formalism account in an elegant way for the interaction of charged particles with the electromagnetic field. In general, fields determine the way particles move, while particles can also make fields to vary, a statement that is beautifully revealed in electromagnetic theory. From the quantum mechanical point of view, fields and particles are closely related concepts. The *Quantum Theory of Particles and Fields* is an active research field in the frontier of physics and it is nice to be already capable to "scratch the surface" of these matters from classical mechanics.

How to introduce particle-field coupling in the classical field theory? A hint comes from the discussion at the end of the previous subsection. Suppose that the coupling of a particle with a field is able to change the system's mass. A trial particle Lagrangian might be,

$$L = -(m + g\Phi)c^2\sqrt{1 - \frac{v^2}{c^2}}, \tag{4.18}$$

in which g is called *coupling constant*. It measures the intensity of the action of Φ on the particle.

Let us try an interpretation of this Lagrangian by considering the $v \ll c$ limit, keeping only the first term on the $(\frac{v}{c})$ expansion. We find,

$$L = -mc^2 + \frac{1}{2}mv^2 - g\Phi + g\Phi\frac{v^2}{2c^2}. \tag{4.19}$$

The first term is a harmless constant while the last is much smaller than the previous. A good surprise comes when we consider the two terms in the middle, $\frac{1}{2}mv^2 - g\Phi$, which looks like our well known $L = T - V$. That is, when we start from Lagrangian (4.19) , which asserts the field a "mass", for slow particles we recover the scenario of a interaction described by the potential energy, the role played by $g\Phi(t, x)$. Lagrangian (4.19) is a way particle and field couples to each other. If you think, for example, of a particle in Earth's gravitational field, then $g = m$.

Except for the detail that the field can be changed by the particle (recall that the movement of a charged particle creates a magnetic field), we are in position to write a Lagrangian for the whole system, particle plus field. Frequently the full Lagrangian is taken as the sum of the free particle's Lagrangian (its kinetic energy T), that for the field, Eq. (4.13) and that for the interaction $-g\Phi$. But there is still a problem, since L and L represent different things. While the action for the particle is an integral over t, for the field it is an integral over (t, x). The way to put them together is to rewrite L using a Dirac delta function, so that the interaction term becomes $-g\delta(x - a)\Phi(x, t)$, where a is the position of the particle. With this artifice, we can now write the action for the particle-field system,

$$S = \int_a^b [\frac{1}{2}(\frac{\partial \Phi}{\partial t})^2 - \frac{1}{2}(\frac{\partial \Phi}{\partial x})^2 - g\delta(x)\Phi + \frac{g\dot{x}^2}{2c^2}]dxdt. \tag{4.20}$$

Recovering the other space terms in (y, z), Lagrange's equation (4.10) becomes,

$$-\frac{\partial^2 \Phi}{\partial t^2} + \nabla^2 \Phi = g\delta(x - a(t)), \tag{4.21}$$

where $a(t)$ accounts for the particle's motion, which will affect the field itself. This is a very complicated field equation! We can, however, get a clue about it by considering a static particle. In this situation, the above equation becomes,

$$\nabla^2 \Phi = g\delta(x - a), \tag{4.22}$$

which resembles Poisson's equation of electromagnetism.[5]

4.5   Final Considerations on Field Theory

This seems to be a good point to finish the chapter, for it is rather a primer on field theory, aiming mainly at showing that Hamilton's principle naturally embodies the dynamics of fields, not only of particles. As the field becomes the very degree of freedom, the space coordinates assume a role analogous to that time t plays in the particle Lagrangians. Some advances towards the relativity concepts are then naturally introduced to build a Lagrangian theory of even non-relativistic fields, as the treatment of space and time being taken on the same footing. However, it is the Lorentz invariance of the proper time the real basis for constructing a relativistic field Lagrangian. Space and time becomes a unique thing, space-time, with profound physical and philosophical implications, while the coincidences of the forms of the non-relativistic and relativistic Lagrangians is one more example of those outstanding things we met when doing physics.

---

[5] It is not however, because the electromagnetic field is a vector field. To see how it is properly treated, refer to the book by Susskind and Friedman in the bibliography.

Bibliography

1. J. Spandaw, Apparent failure of the principle of least action. Am. J. Phys. **81**, 144 (2013)
2. K.S. Krane, The falling raindrop: Variations on a theme of Newton. Am. J. Phys. **49**, 113 (1981)
3. R.P. Feynman, Space-time approach to non-relativistic quantum mechanics. Rev. Mod. Phys. **20**, 367 (1948)
4. E. Nöther, Invariante Variationsprobleme. Nachr. Gesell. Wissenchaft. Göttingen **2**, 235 (1918)
5. A.S. de Castro, Exploring a rheonomic system. Eur. J. Phys. **21**, 23–26 (2000)
6. J.R. Mohallem, Galilean invariance in Lagrange mechanics. Am. J. Phys. **83**, 857 (2015)
7. H. Hartmann, Die bewegung eines körpers in einem ringförmigen potentialfeld. Theor. Chim. Acta. **24**, 201 (1972)
8. L.A. Poveda, M.A. Muniz, J.R. Mohallem, Chaos and resonances in the classical scattering of a positron by a model diatomic molecule. J. Mol. Model. **29**, 65 (2023)
9. J. Mitroy, I.A. Ivanov, Semi-empirical model for positron scattering and annihilation. Phys. Rev. A **65**, 04270 (2002)
10. J.R. Mohallem, Two helpful developments towards better understanding of adiabatic invariance in classical mechanics. Rev. Bras. Ens. Fis. **41**, e20180214 (2019)
11. F.S. Crawford, Elementary examples of adiabatic invariance. Am. J. Phys. **58**, 337 (1990)
12. V.I. Arnold, *Mathematical Methods of Classical Mechanics* (MIR, 1987)
13. G.R. Fowles, G.L. Cassiday, *AnalyticalMechanics* (Thomson Brooks/Cole, 2005)
14. H. Goldstein, *Classical Mechanics* (Addison-Wesley, 1980)
15. L. Landau, E. Lifchitz, *Mechanics* (MIR, 1966)
16. N. Lemos, *Mecanica Analitica* (Livraria da Fisica, 2007)
17. A. Sommerfeld, *Mechanics* (Academic Press, 1952)
18. L. Susskind, A. Friedman, *Special Relativity and Classical Field Theory - The Theoretical Minimum* (Penguin Books, 2017)
19. S.T. Thornton, J.B. Marion, *Classical Dynamics of Particles and Systems* (Brooks/Cole, 2004)

Index

© The Editor(s) (if applicable) and The Author(s), under exclusive license to Springer Nature Switzerland AG 2024
J. R. Mohallem, *Lagrangian and Hamiltonian Mechanics*,
https://doi.org/10.1007/978-3-031-55202-1

MIX
Papier aus verantwortungsvollen Quellen
Paper from responsible sources
FSC® C105338

If you have any concerns about our products,
you can contact us on
ProductSafety@springernature.com

In case Publisher is established outside the EU,
the EU authorized representative is:
Springer Nature Customer Service Center GmbH
Europaplatz 3, 69115 Heidelberg, Germany

Printed by Libri Plureos GmbH
in Hamburg, Germany